maths
for advanced **biology**

Alan Cadogan and
Malcolm Ingram

Published in 2002 by:
Nelson Thornes Ltd
Delta Place
27 Bath Road
CHELTENHAM
GL53 7TH
United Kingdom

02 03 04 05 06 / 10 9 8 7 6 5 4 3 2 1

A catalogue record for this book is available from the British Library

ISBN 0 7487 6506 9

Illustrations by Oxford Designers and Illustrators
Page make-up by Mathematical Composition Setters Ltd

Printed and bound in Croatia by Zrinski

Contents

Preface   v

1 Numbers – Egypt to metric   **1**
1.1 Numbers and the biologist   1
1.2 Basic forms of notation   2
1.3 Rounding and significant figures   9
1.4 Using a scientific calculator   11
1.5 Use of tables in biology   12
Exam questions   13
Answers to Test Yourself Questions   14

2 Symbols and units – the Greeks have a word for it   **15**
2.1 Words and the biologist   15
2.2 Using symbols for size in biology   16
2.3 Using scales and measuring biological objects   17
2.4 Using units and symbols   19
2.5 Units   28
Exam questions   33
Answers to Test Yourself Questions   34

3 Algebra – x – the unknown   **36**
3.1 Equations and rules   36
3.2 Equations and ecology   40
3.3 The Hardy–Weinberg equilibrium   41
Exam questions   42
Answers to Test Yourself Questions   43

4 Indices, powers of ten, scientific notation   **44**
4.1 Indices   44
4.2 Powers of ten   45
4.3 Scientific notation (or standard form)   48
Exam questions   49
Answers to Test Yourself Questions   50

5 Ratios and getting things in proportion   **51**
5.1 Ratios   51
5.2 The use of ratios in genetics   52
5.3 Ratios in other biological examples   56
Exam questions   61
Answers to Test Yourself Questions   61

6 Displaying data: graphs, charts and scales   **62**
6.1 Dealing with data   62
6.2 Drawing graphs   67
6.3 Plotting and drawing line graphs   75
6.4 Other types of graph   79
Exam questions   81
Answers to Test Yourself Questions   83

7 Using graphs and interpreting data   **85**
   7.1 Interpreting graphs   85
   7.2 Calculations from graphs   86
   7.3 Logarithmic scales   89
   7.4 Using graphs   91
   Exam questions   98
   Answers to Test Yourself Questions   100

8 Correlation and regression   **101**
   8.1 Correlation   101
   8.2 Drawing lines of best fit   105
   8.3 The accurate way of drawing the regression line   108
   8.4 A correlation coefficient for ecologists   110
   Exam questions   112
   Answers to Test Yourself Questions   113

9 What does 'mean' mean?   **114**
   9.1 Averages   114
   9.2 Describing population and samples   118
   Exam questions   121
   Answers to Test Yourself Questions   122

10 Distribution   **123**
   10.1 Standard deviation   123
   10.2 Confidence   131
   Exam questions   132
   Answers to Test Yourself Questions   133

11 Chi-squared: a test of closeness   **134**
   11.1 Chi-squared test and the null hypothesis   134
   Exam questions   141
   Answers to Test Yourself Questions   142

12 *t*-test, *U*-test and *D*-test   **143**
   12.1 Comparing samples – the 'Student's *t*-test'   143
   12.2 The Mann–Whitney *U*-test   146
   12.3 The Simpson's diversity index, *D*   149
   Exam questions   150
   Answers to Test Yourself Questions   152

Appendix   **153**
Answers to exam-type questions   **154**
Further practice questions   **156**
Index   **160**
Acknowledgements   **162**

Preface

As courses and examinations in biology have changed, students have been expected to have acquired more skills and use more techniques. It is now much more important to understand how scientists write and in the early days of learning biology people are now expected to be able to communicate science ideas with others.

Precision is the norm in science and this can only be achieved with good mathematical skills. We do not believe that you have to be a *great* mathematician to study Biology. All too often students are frightened away from studying the sciences because of an irrational fear of numbers. Our aim in the present text is to show how anyone can master the essential number work in a study of biology at this level.

We have tried to do this by assuming nothing about the reader's previous experiences in arithmetic and algebra. We start with basics and remind you about some of the things you probably first came across years ago! Topics such as rounding-up, squares and square roots, ratios and percentages are touched upon. There is some concentration on statistical topics because it is now expected that you will use a suitable statistical technique in most reports of practical investigations. Our examples use simple numbers because it is easier to learn the process that way. Once you know the process you can apply it to your real data.

Although, between us, we have a great deal of experience teaching and examining students at this level and at graduate level we have consulted widely with colleagues currently examining with other Awarding Bodies. The reports of data collected come from the work of a range of students. We are grateful to them and to all that we have consulted.

Maurice Wilkins (the 1962 Nobel prize winner with Francis Crick and James Watson) wrote: '*Biological education will continue to have an importance in its own right, but it will also be very important as part of a broader education, not simply for science graduates, but for graduates of all kinds ... One will realise that man is part of nature, rather than being above it.*'

Charles Darwin, one of the greatest biologists of all time felt inhibited by his failure to study maths. He wrote, '*I deeply regretted that I did not proceed far enough to at least understand something of the great leading principles of mathematics.*'

We have used the Institute of Biology *Biological nomenclature (3rd Edition) 2000* and also referred to the Association of Science Education *Signs, Symbols and Systematics 2000*.

We have listed all the maths topics (from arithmetic to statistics) that your course needs. Each topic has a short chapter that has

● a list of objectives for you to achieve

● an introduction to put the topic into context

● an explanation – with real examples

● some 'Test Yourself' questions

● the answers to these – at the end of the chapter

● some exam-type questions at the end of each chapter with the answers to these at the end of the book.

To help you to find your way around the book, we have used a simple code.

Test Yourself

These are in the text and you should try each one as you come to it. If you can do it, you have understood the point. If you cannot, it means that you need to think again about the text above the symbol.

shows a section on using a calculator. We suggest that it should be of the VPAM (visually perfect algebraic method) type. These show you the sum you have entered as well as the result. It is worth learning how to use with confidence and we give some hints on technique.

HINT

offeres some help for dealing with a problem. This might be a short cut or a reminder of a point of technique.

means that the data has been collected by students in the field.

means that the data have been collected in the laboratory.

Exam Questions

means a question of the sort you may find in an examination. Try all of them! Practice makes perfect.
We have only given the numerical parts of the answers to these questions. Further details will be on the web.

About the calculator

We have assumed that you will use a VPAM calculator. All our instructions are for the Casio *fx*-83WA which is representative of most. If you have a different calculator, find out from its manual how to perform each operation.

About units

We have used SI units throughout, with all the usual conventions for abbreviations and unit names.

Thanks

We want to thank Beth Hutchins at Nelson Thornes for her help and advice and to express our appreciation to the Education Committee of the Royal Statistical Society for their helpful comments on an early draft of the text.

Alan Cadogan
Malcolm Ingram

Chapter 1

Numbers – Egypt to metric

After completing this chapter you should:

- have revised basic calculations, fractions, decimal notation and percentages
- be able to give rounded answers and significant figures
- have considered making estimates without the use of a calculator
- have explored using a scientific calculator to find x^n, $\frac{1}{x}$, $\log_{10} x$ and $\sqrt{x}$
- have examined the use of tables in biology.

1.1 Numbers and the biologist

It is only in fairly recent times that mathematical calculations have been seen as an important part of biological work. Gregor Mendel, the Czech monk who was the 'father of genetics', had problems in 1865 getting his work accepted. There was an 'anti-mathematical' prejudice amongst the biologists of the day that continued even after Mendel's work was re-discovered in 1900. The writer of another biological paper in 1900 was told by the Royal Society that statistics should not be used because 'mathematics should be kept apart from biological application'.

Fig. 1 *Gregor Mendel (1822–84) postage stamp to mark the centenary of his original publication*

A century later it seems ludicrous that such attitudes could have existed. Most new research in life sciences is published together with strong supporting mathematical evidence. Research studies in ecology quote the evidence of investigations by displaying tables and graphs and inevitably test the evidence by statistical methods. Discoveries in medicine, food science, biotechnology, agriculture and pharmacy would not be accepted for publication today unless the science writer presented data in a form that had been tested and could be re-tested by other scientists (see *Figs 2* and *3*).

It isn't just a case of biologists becoming mathematicians but of them using mathematical techniques as a tool in their investigations. It has become inevitable that students studying biology today need to have some mathematics to understand certain areas of the subject and to be able to write proper scientific reports of their projects and investigations. However, present day students have access to simple calculators for straightforward arithmetic and inexpensive scientific calculators that, with the help of the manual, can be made to perform any calculations needed.

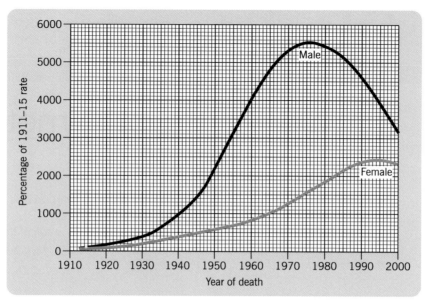

Fig. 2 *Part of a medical report showing deaths from lung cancer (ages 0–84)*

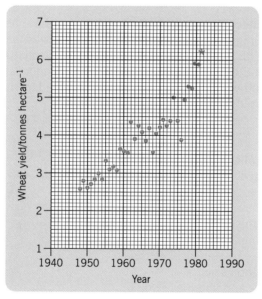

Fig. 3 *Graph showing increase in crop yield of wheat in UK*

1.2 Basic forms of notation

Five thousand five hundred years ago the Egyptian mathematicians had a system of number **notation** that worked well and can still be used today. It was based on hieroglyphs: using a symbol for a staff as number one, a heel print as ten, a coiled rope as 100, a lotus plant for 1000, a bent finger for 10 000 and a frog or tadpole for 100 000 (biological mathematicians even at that time?).

Biology students today are not expected to be familiar with that system of notation, or even the Roman one that you will know from clock- and watch-faces: (1 = I, 5 = V, 10 = X, 100 = C and 1000 = M). Our system of notation is based on the **decimal system**. We count in tens

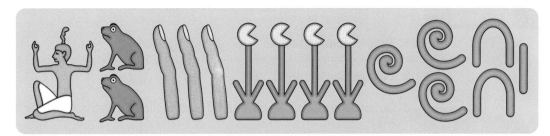

Fig. 4 *Egyptian hieroglyph for the number 1 234 321 – the kneeling genie represents one million*

and scientists use the metric system and the International System of Units (**SI**). It is strange that scientists in the United States are unique in the developed world in resisting the adoption of SI (see also p. 28).

Up to now you will certainly have used a great deal of maths notation – it is really like speaking the language of 'maths'. If you see printed: 1 + 3 = 4 you will say it as the words one plus three equals four. You will come across signs and symbols that are special to that language. In Chapter 2 we explain some of these symbols and later in this chapter the abbreviations d.p. and s.f. are used. In a way, this book is all about mathematical notation and it is now time to look into some types of basic notation – fractions, decimals and percentages.

Fractions and decimal notation

(This part is just to remind you of some of the simple rules of the maths language.)
In some languages you must read across the page, either left to right or right to left; other languages are read vertically. Often in maths notation we have to work in *all* directions. This is particularly true of calculations involving fractions.

Example

$$\frac{2}{3} + \frac{1}{4} = \frac{11}{12}$$

When you have a fraction like $\frac{2}{3}$, you can present it in different ways by multiplying both the top and the bottom by the same number. So $\frac{2}{3}$ is the same as $\frac{8}{12}$ and $\frac{1}{4}$ is the same as $\frac{3}{12}$.

HINT *This easy example is a reminder to you that when thinking about the rules to apply to a complex calculation, you should try it out first using simple numbers such as those used here.*

Example
When writing fractions, try to cancel so that you get the simplest number.

30 out of a group of 45 students play for teams. What fraction is this?
Well, it could be written

$$\frac{30}{45} \text{ or } \frac{10}{15} \text{ or } \frac{6}{9}$$

These are all correct, but the simplest fraction to use is obtained by cancelling down to $\frac{2}{3}$.

Test Yourself   Exercise 1.2.1

1 Simplify these fractions: (a) $\frac{7}{21}$ (b) $\frac{13}{52}$ (c) $\frac{144}{480}$

2 Calculate: (a) $\frac{4}{19} + \frac{6}{19}$ (b) $\frac{1}{3} + \frac{4}{7}$ (c) $\frac{9}{11} - \frac{6}{11}$

3 Out of 80 students, 70 have used a scientific calculator. What fraction of the group has not?

In all of the fractions so far, the top number (numerator) has been smaller than the bottom number (denominator). These are called **proper fractions**. If the top number is bigger than the bottom, like $\frac{5}{3}$, it is called an **improper fraction**. It may be suitable to write it just like that, or you could convert it to a **mixed number** just by dividing the top by the bottom and noting the remainder. In this case $1\frac{2}{3}$.

Test Yourself   Exercise 1.2.2

1 Express the following mixed numbers as fractions: (a) $3\frac{1}{3}$ (b) $2\frac{7}{9}$ (c) $21\frac{1}{4}$

2 Express the following fractions as mixed numbers: (a) $\frac{34}{3}$ (b) $\frac{19}{8}$ (c) $\frac{510}{100}$

Multiplying and dividing fractions
In order to multiply fractions you should first, if necessary, convert any mixed numbers to fractions.

Example
To multiply $\frac{1}{3}$ by $\frac{3}{4}$:
Multiply together the top numbers:

$1 \times 3 = 3$

and the bottom numbers:

$4 \times 3 = 12$

They give the new fraction $\frac{3}{12}$ which equals $\frac{1}{4}$.

To divide, again convert any mixed numbers to fractions.

Example
$\frac{2}{3} \div \frac{1}{6}$

Change the ÷ sign to × and turn the fraction after the sign upside-down.
So you can rewrite it as: $\frac{2}{3} \times \frac{6}{1}$
When you then multiply it becomes $\frac{12}{3}$. So the answer is 4.

Test Yourself   Exercise 1.2.3

1 Calculate: (a) $\frac{2}{3} \times \frac{5}{6}$ (b) $4\frac{1}{2} \times 1\frac{1}{4}$

2 Calculate: (a) $1\frac{1}{3} \div \frac{8}{33}$ (b) $1\frac{4}{5} \div 1\frac{1}{2}$

Decimals and fractions

You need to know how to convert fractions to decimals and decimals to fractions. The first is easy.

Example

The fraction $\frac{9}{5}$ is worked out just by dividing 9 by 5 (either the long way or by keying into your calculator $\boxed{9}\,\boxed{\div}\,\boxed{5}\,\boxed{=}$) to get the answer 1.8.

Example

To convert a decimal number such as 2.25 into a fraction you have to remember that the first 2 is a whole number, that the 2 after the decimal point represents two tenths and the 5 is five hundredths. So the number can be written:

$$\begin{array}{ccc} 2 & 2 & 5 \end{array}$$
$$2 + \tfrac{2}{10} + \tfrac{5}{100} \quad \text{or } 2 + \tfrac{20}{100} + \tfrac{5}{100} = 2 + \tfrac{25}{100}$$

Then by cancelling you get $2\frac{1}{4}$.

Percentages

Newspapers and other media tell us about percentages almost every day. We hear that the cost of living has increased by $x\%$, or that we have had $y\%$ of the month's rainfall, or that we scored $z\%$ in last week's test. We all know what the term means – or do we? Biologists often have to use percentage calculations to express results of an investigation in an accurate and understandable way.

Example

As a simple case study, I looked at my garden soil. It had always been hard to dig and I knew that it was a heavy clay soil; I decided to do a simple analysis. I took a series of samples, each of about 25 g, and allowed the soil to dry completely for several days. The total mass then was 327 g. I knew that I could heat it to burn off the organic material; after doing so there was only 265 g remaining. I crushed this and emptied it into a sieve with a mesh of 0.5 mm. I noted that I had 49 g left on the sieve and that 216 g had passed through it.

Restaurant	Table No.46
Soup	3.00
Main course	6.00
Pudding	3.00
V.A.T. 17.5% Service 10%	
TOTAL	16.00

Fig. 5 'Was the waiter correct?'

I had learned something from this investigation, but certainly didn't have results that could be expressed in a scientific way. I was also unable to compare my results with the soil in my colleague's garden. So I wrote out the result in a table.

Table 1 *Analysis of garden soils*

	Organic matter/g	Inorganic particles (less than 0.5 mm)/g	Inorganic particles (more than 0.5 mm)/g
Garden 1	62	216	49
Garden 2	53	185	165

How could you improve Table 1? First check for Garden 1 that it all adds up to the original mass of the sample i.e.

62 g + 216 g + 49 g = 327 g

Now, deal with each component: starting with the organic material in the soil – it is 62 g in a total mass of 327 g dry soil.

Per cent just means 'for every hundred' – we know that we have 62 g per 327 g. We need to convert this to 'x per 100 g'.

HINT

You might already have spotted, even without a calculator, that it is approximately 60 per 300 or 20 per 100. So we have a rough estimate of 20%.

However, to work it out accurately it is:

62 g ÷ 327 g × 100% = 18.96%

6 2 ÷ 3 2 7 × 1 0 0 = or 6 2 ÷ 3 2 7 shift %

Similarly, to calculate the percentage of inorganic particles less than 0.5 mm.

$$\frac{216 \text{ g}}{327 \text{ g}} \times 100\% = 66.06\%$$

The percentage of particles bigger than 0.5 mm is

$$\frac{49}{327} \times 100\% = 14.98\%$$

These percentage results can now be entered in Table 2.

Table 2 *Analysis of garden soil expressed as percentages*

	Organic matter/%	Inorganic particles (less than 0.5 mm)/%	Inorganic particles (more than 0.5 mm)/%
Garden 1	18.96	66.06	14.98
Garden 2			

Test Yourself

Exercise 1.2.4

First check that the percentages recorded in Table 2 add up to 100%. You will sometimes have to accept that there will be a *small* departure from 100%.

1 Calculate the data for Garden 2 so that you can complete Table 2.

HINT

Start by working out the total mass of dry soil used from Table 1.

2 Later (after reading p. 79) you could show this data as a pie chart.

The above type of calculation is sometimes needed in questions about nutrition.

Example

The label on a jar of powdered coffee creamer has this nutrition information:

	Amount per 6.5 g spoonful/g	%
Protein	0.08	
Carbohydrate	3.90	
Fat	2.00	

What percentage of the powder is fat?

The fraction of the powder that is fat is

$$\frac{2.00 \text{ g}}{6.50 \text{ g}}$$

This is expressed as a percentage by multiplying by 100%.
So the amount of fat in the creamer is:

$$\frac{2.00 \text{ g}}{6.50 \text{ g}} \times 100\% = 30.77\%$$

Do the same for the carbohydrate and protein.
The amount of carbohydrate is:

$$\frac{3.90 \text{ g}}{6.50 \text{ g}} \times 100\% = 60.00\%$$

The protein is:

$$\frac{0.08 \text{ g}}{6.50 \text{ g}} \times 100\% = 1.23\%$$

> **HINT**
>
> *It is worth remembering to write the units each time through a calculation. Even in the easy examples above, the units (g) above and below the line cancel each other out so that for the answer you are left only with the %. Such questions in an exam give an opportunity to pick up a few marks without too much effort.*

Test Yourself

Exercise 1.2.5

The table below gives the mean values for *body mass* and *lean body mass* for two groups of women. One group is of lean women and the other is of moderately obese women.

Group	Body mass/kg	Lean body mass/kg
Lean women	56	41
Moderately obese women	88	49

1 Calculate the percentage of lean tissue and also the percentage of adipose tissue (fat) in each group.

> **HINT**
>
> *Adipose tissue = body mass – lean body mass*

(Data from Edexcel, GCE Biology and Human Biology, Jan 1997, Module B/HB4D)

Figure 6 shows the record of the most common causes of death at St Botolph's, London, 1583–99.

In 1988 infant mortality varied in different developing countries. In some, such as Jamaica (1.1%), Thailand (3.0%) and Sri Lanka (3.2%) it was fairly low. However in Saudi Arabia it was 7.0%, Morocco 8.0% and in the Ivory Coast as high as 9.5%

(Data from WHO and UNICEF).

Plague	23.6%
Consumption (and Convulsion)	22.2%
(not stated)	14.1%
Pining, decline	13.2%
Ague, fever	6.1%
Flux, colic	2.5%
Smallpox	2.4%
Childbed	1.5%

Fig. 6 *The most common causes of death – St Botolph's, London, 1583–99*

It is impossible to study human health and disease statistics without using percentages. To calculate the total number of deaths per year in the WHO and UNICEF statistics you need to know the total number of babies born in each of the countries.

Some questions are set using not 'per hundred' but 'per thousand' or some other number, such as in the following example.

Example
In one year the death rate from influenza in the United States was 0.3 per 100 000. The population of the United States was about 250 million. How many people died of influenza in this year? Show your working.

(Data from London GCE Biology, B2, June 1997)

The calculation is of the same sort as a percentage one:

$$\frac{0.3}{100\,000} \times 250\,000\,000 = 0.3 \times 2500 = 750 \text{ people}$$

Percentage change
The other type of percentage calculation you may have to perform is to find a percentage change.

Example
If an American aunt sent you $45 as a birthday gift and promised a 10% increase for your next birthday, you could probably work out that 10% of $45 is $4.50 and anticipate a cheque for $49.50.

Example
What about the population of earthworms – found to be 63 per square metre at one survey and 81 per square metre the following year?

There had been an increase of 18 earthworms. To find the percentage increase we work out:

Change ÷ original population × 100%

So

$18 \div 63 \times 100\% = 28.57\%$

The percentage increase in the earthworm population is 28.57%.

1 Examine the data in the table below and calculate the estimated total population of lapwings breeding in England and Wales in 1998.

2 Select the region with the biggest percentage decrease and the region with the smallest percentage decrease and calculate the change in number in the populations of those regions.

UK Region	1987 Estimated population	% population change (1987–98)
North	29 517	−48
South-east	9567	−46
South-west	5741	−71
Wales	7550	−78
Yorkshire	24 248	−33
Total UK	121 957	−48

(Data adapted from British Trust for Ornithology Census, 1999)

1.3 Rounding and significant figures

Some students make the mistake of giving an answer to a degree of accuracy that is really impossible. They may measure the height of three colleagues at 175 cm, 161 cm and 154 cm. Then, using the calculator, they may report that the mean height is 163.333 33 cm. This suggests that it is a very accurate answer. In fact, since the three subjects were measured to within the nearest centimetre the answer should be given by **rounding** to the nearest whole number. In this case it would be to 163 cm. However, if on remeasuring the first colleague he was actually found to be 176 cm, the new mean value would be 163.666 66 cm which would have to be rounded *up* to 164 cm.

KEY FACT   *The rule is that 0.5 and over gets rounded up and below 0.5 gets rounded down.*

The same rule (0.5 and over rounds up) can be applied to other degrees of accuracy. Suppose that you get an answer to a calculation that is 863.6518 g. If you are asked to give the answer to the **nearest gram**, it would be 864 g because you would have rounded up the final 3.6 to 4. Giving the answer to the nearest 100 g would mean to rounding up to 900 g.

Similarly, using this same answer but to one **decimal place** would round 863.65 g up to 863.7 g. Expressing 863.6518 g to two decimal places would give 863.65 g. To explain what you have done, it is sometimes advisable to state that the answer is given to a number of decimal places, e.g. 863.65 g (2 d.p.). (Incidentally each of the three people measured earlier is, to the nearest metre, 2 m tall!)

Rounding up and rounding down to a defined degree of accuracy is a way of producing significant figures in an answer. Scientists usually have to measure some property and after working through a calculation they have to give a result. Say you timed a jaguar running over a distance of 65 m and recorded a time of 2.43 s. You could work out with your calculator that it was travelling at 26.748 97 m s^{-1}. How many of these decimal places should you include in the answer? Well, the convention is that you count the number of **significant figures** in each measurement that went into the calculation (i.e. 65 m – two significant figures and 2.43 s – three significant figures) and use the smaller number for the number of significant figures in the answer. So the answer would be 27 m s^{-1} (to 2 s.f.). So that is the first rule for quoting significant figures and it is applied whenever you are multiplying or dividing.

If adding or subtracting numbers, the rule that applies to decimal places is that in the final answer the same number of decimal places should be given as is found in the measurement with the fewest number of decimal places.

Example
If a 7.1 kg dog carries a 0.35 kg bone, the combined mass could be calculated as 7.45 kg – but since the mass of the dog was only given to one decimal place, the same treatment should be given to the total. To one decimal place, the answer should be 7.5 kg – remember the rule about rounding up?

Estimating results without using a calculator

In the garden soil calculation (p. 6) it was suggested that you should do a rough estimate to make sure that the answer that you get is in the correct range. (I made a mistake once in a calculation and worked out that a tall tree was 0.35 m high! A quick visual check told me that it was about 30 m tall. I was out by a hundred times. This 10^2 can be thought of as *two* orders of magnitude. I had made a mistake but it was easy to see and to correct.)

Often these mistakes are less obvious. It is always worth doing a rough estimate before you go through a calculation – particularly with a calculator, where it is easy to put a decimal point in the wrong place. You could then be out in your answer by one, two or three orders of magnitude – either above or below the actual answer.

Example

9.8 × 697.0 ÷ 102.4

Make a rough estimate. Is the answer likely to be about 7, or 70, or 700, or 7000 – what do you think? The method to be used is to round off the numbers involved, cancel out and approximate.

You probably worked out: 10 × 700 ÷ 100 is about 70. Well, calculation gives the answer 66.7 – so the estimate tells us that this is likely to be correct. So we now have only to realise that one of the numbers involved (9.8) had only two significant figures and therefore a reasonable rounded-up answer would be 67 (2 s.f.).

In making estimates it is very important to really understand powers of ten – but that is the subject of a later chapter (see p. 45).

1.4 Using a scientific calculator

One of the reasons for having a scientific calculator, rather than the sort that would be used for checking the shopping, is that you are able to undertake special calculations. There are four that you may need to use. They are:

- to find x^n
- to find $\dfrac{1}{x}$
- to find $\log_{10} x$
- to find $\sqrt{x}$

How to find x^n

On p. 44 there is an explanation that this symbol can be read as 'x raised to the power n' or simply as 'x to the n'. So 5^2 is 'five to the power 2' (actually in this case we usually say 'five squared').

Check your calculator to see if it has a key:

$\boxed{x^y}$

If so (you need to think of 5 as x and 2 as y), then the keys that you use are: $\boxed{5}\,\boxed{x^y}\,\boxed{2}\,\boxed{=}$ and you should read the answer 25.

In other words, the number 5 multiplied by itself, 5^2, is the same as $5 \times 5 = 25$.

You can try this out with other numbers: what about 9^2 and 9^3?

How to find $\dfrac{1}{x}$

If x is the number, then $\dfrac{1}{x}$ (read as 'one over x') is described as the reciprocal of the number.

So $\frac{1}{5}$ is the **reciprocal** of 5 and obviously is calculated to be 0.2.

Sometimes you may have to plot a graph using the *reciprocals of the measurements* instead of the measurements recorded.

Example

If your results were, in seconds, 5, 8, 11 and 12.5, then the reciprocals to be plotted would be 0.2, 0.125, 0.09 and 0.08. This is how we would key for the reciprocal of 12.5:

$\boxed{1}\,\boxed{\div}\,\boxed{1}\,\boxed{2}\,\boxed{.}\,\boxed{5}\,\boxed{=}$

However, there is one key that simplifies this; it is $\boxed{x^{-1}}$.
So the keys to use to get the same result are:

$\boxed{1}\,\boxed{2}\,\boxed{.}\,\boxed{5}\,\boxed{x^{-1}}\,\boxed{=}$

Now *check* the other reciprocals of 5, 8, and 11.

How to find log₁₀ x

Your calculator should have one key: [log]

This will give you the 'log to the base ten' of any number. So if you want to find $\log_{10} 8$, you simply key in: [log] [8] [=] You should get 0.9031.

(With some calculators you may have to key in: [8] [log] [=].)

This calculation may be needed if you have to calculate the pH of a solution. It is also a technique that is used to re-plot graph points when a logarithmic scale may separate them and make it easier to read the graph – sometimes as a straight line (see p. 89).

HINT | *To get back from the log to the base ten you key in:* [shift] [10ˣ] [=] *and the 0.9031 goes back to 8.0.*

How to find √x

To find the **square root** of a number is like reversing the squaring of a number. This means that $\sqrt{25} = 5$ and $\sqrt{9} = 3$. Your calculator probably has a [√] key. If so, the keys to use for $\sqrt{25}$ are probably: [√] [2] [5] [=]

Some of these techniques will be demonstrated later with examples from biological calculations.

1.5 Use of tables in biology

Finally in this chapter it is necessary to discuss the use of tables. We have already used several; it is a technique of showing information clearly. You will be expected as a trained scientist to be able to refer to tables of data and to spot trends and patterns. When some people look at text that includes tables, their eyes just skip over the table and come back to focus on the text. They do not bother to extract from the numbers as much information as they could. Take the following example:

Example

Table 3 *Heat and water produced from different food types*

	Heat produced in calorimeter/kJ	Water produced on oxidation/g
1 g protein	23.4	0.41
1 g carbohydrate	17.6	0.55
1 g fat	38.9	1.07

Now, spend some time looking closely at the table for the details:

● Check the headings of the two columns – do you understand what they mean? (Notice that what is measured is separated from the units by a solidus, '/').

● Look at the three items in the first column and note that the same mass of each food type is listed. Now 'ask the table some questions':

1 Which type of food produces least heat per gram in the calorimeter (and hence the lowest energy yield for the body)? (Notice that the units are not given in the columns because the heading to the column gives that information.)

2 Compare this food type with the food type that produces the most energy per gram – approximately how much greater is the yield?

3 Compare the amounts of water produced when equal amounts of the three food types are broken down by oxidation (or respiration in the body)? Does the food type yielding the smallest amount of energy also yield less water than the other two types?

Now we are going to use the data provided in Table 3 above to perform some simple calculations.

Test Yourself

Exercise 1.5.1

1 A 175 g pot of fruit yoghurt is labelled as containing 6.5 g protein, 26.3 g carbohydrate and 6.8 g fat. Use the data in Table 3 to find the total energy value of the yoghurt measured in kilojoules.

2 A bag of easy-cook brown rice weighing 130 g gives the information that 100 g provides 6.9 g protein, 74 g carbohydrate and 2.8 g fat.
A bag of easy-cook white rice also weighs 130 g and every 100 g has 7.4 g protein, 80 g carbohydrate and 0.5 g fat.

An athlete has to decide which type would provide the most energy if equal quantities were eaten. Show the calculation you would use to convince her which to choose.

Exam Questions

Exam type questions to test understanding of chapter 1

1 Dippers are small birds that feed on fish and aquatic insects in shallow rivers. They carry food back to the young nestlings. A survey of the dippers in a river in Wales showed what had been eaten by the adults and their young in terms of the number of items and the energy they obtained from the items eaten.

Table 4 *Diet of adult dippers and nestlings in a Welsh river during a period in 1988*

	Total number of items		Total energy of items/kJ	
	adults	nestlings	adults	nestlings
Fish	10	10	46.95	46.95
Mayfly nymphs	484	269	19.83	11.85
Caddis larvae	176	430	34.18	135.10
Other	144	66	6.36	2.93
Totals	814	775	107.32	196.83

(a) Adult dippers and nestlings each consume 10 fish. Calculate what percentage of number of items and what percentage of total energy is provided by fish to each age group.

(b) Decide which food item gives nestlings the greatest amount of energy. State this as a percentage of the total energy.

Answers to Test Yourself Questions

Exercise 1.2.1, *p.4*

1 (a) $\frac{1}{3}$ (b) $\frac{1}{4}$ (c) $\frac{3}{10}$

2 (a) $\frac{10}{19}$ (b) $\frac{19}{21}$ (c) $\frac{3}{11}$

3 $\frac{70}{80}$ have, so $\frac{10}{80}$ have not, answer $\frac{1}{8}$

Exercise 1.2.2, *p.4*

1 (a) $\frac{10}{3}$ (b) $\frac{25}{9}$ (c) $\frac{85}{4}$

2 (a) $11\frac{1}{3}$ (b) $2\frac{3}{8}$

  (c) $5\frac{1}{10}$ (Did you convert $\frac{10}{100}$ to $\frac{1}{10}$?)

Exercise 1.2.3, *p.4*

1 (a) $\frac{5}{9}$ (b) $5\frac{5}{8}$

2 (a) $5\frac{1}{2}$ (Did you remember to cancel out?) (b) $1\frac{1}{5}$

Exercise 1.2.4, *p.6*

1 Organic matter = 13.15%; Inorganic less than 0.5 mm = 45.91%; Inorganic more than 0.5 mm = 40.94%

Exercise 1.2.5, *p.7*

1 Lean women: 73% lean tissue and 27% adipose tissue
Moderately obese women: 56% lean tissue and 44% adipose tissue

Exercise 1.2.6, *p.9*

1 Total population in 1988 = 63 418

2 Biggest percentage decrease – Wales: change is from 7550 to 1661 = 5889
Smallest percentage decrease – Yorkshire: 24 248 to 16 246 = 8002

Exercise 1.5.1, *p.13*

1 879.5 kJ

2 Brown = 1573 kJ 100 g^{-1} and white = 1601 kJ 100 g^{-1}; she should choose white for energy

Chapter 2

Symbols and units – the Greeks have a word for it

After completing this chapter you should be able to:

- *use some of the symbols that are common in biological science*
- *understand the symbols used to measure macroscopic and microscopic objects*
- *calibrate a microscope*
- *understand scales used in microscopy and on photographs*
- *explain SI units, base and derived*
- *understand derived (compound) units.*

2.1 Words and the biologist

Students who study biology often comment that there are so many words to learn, it seems like a new language! Well in fact they are partly correct, in that many of the terms used are derived from Greek or Latin; but you don't need to be a classical scholar to learn them. Some parts of biological words come from a handful of Greek and Latin words. It is not essential to know them, but it sometimes helps in understanding the meaning of technical terms. There are patterns and, once you understand a part of a word, you can make assumptions about the meaning. Table 1 shows a few examples:

Table 1 *Some common word roots*

Word root	Meaning and example
aer-	*air/atmosphere*: as in aerobic (meaning 'with oxygen')
amphi-	*both*: as in amphibian (both land and water), amphoteric (both base and acid functions) and some molecules are known as amphipathic (both hydrophobic and hydrophilic elements)
cyt- (also *-cyte* or *cyto-*)	*cell*: as in cytoplasm (cell contents), phagocyte (a microbe-engulfing blood cell) and cytology (the study of cells)
hyper-	*over/excess*: used in describing a higher water potential of a cell or solution, for example, hypertonic and also hypertension
quadr-	*four*: quadrat, quadratic and also a quadruped (which has four legs)

We use some Greek letters as symbols. To help you to avoid the problems that students often have of trying to describe these 'Greek squiggles', Table 2 shows the letters that are commonly used by scientists – though only a few are used at this level in biology. You will soon become familiar with the ones that you need to use. One of these is used in the next section.

Table 2 *The Greek alphabet – note that scientists only use certain letters. Those shown here in bold print are in common usage at this level*

α	**alpha**	ν	nu
β	**beta**	ξ	xi
γ	**gamma**	o	omicron
$\Delta\ \delta$	**delta**	π	**pi**
ε	epsilon	ρ	**rho**
ζ	zeta	$\Sigma\ \sigma$ or ς	**sigma**
η	eta	τ	tau
$\Theta\ \theta$	theta	υ	upsilon
ι	iota	ϕ	phi
κ	kappa	$X\ \chi$	**chi**
$\Lambda\ \lambda$	**lambda**	ψ	**psi**
μ	**mu**	$\Omega\ \omega$	omega

2.2 Using symbols for size in biology

Now look at some common terms used in biology:

microscope, micrometer, micropyle, micropropagation, microhabitat

All these words start with 'micro-'. This prefix denotes 'very small size'. It is also known by its symbol µ, which is a Greek letter (mu, pronounced 'mew').
You can see in Table 2 that sometimes the upper case (capital) letters are used and sometimes the lower case (small) letters.

As well as the prefix 'micro-', other prefixes are used specifically in measuring; it is very important to know them (see Table 3, p. 17).

You will be familiar with **SI** (*Système International d'Unités*) and the SI units of length – metres, centimetres and millimetres. You will have used them for a long time and you will also know that the metre is the **base unit** of length. You can measure the length of a table in centimetres or a bus in metres. However, the largest biological cell is only a fraction of a millimetre and some parts of animal cells are thousands of times smaller than that. Missing from the list in Exercise 2.2.1 is one prefix 'Mega-' (which comes from the Greek *megas* meaning 'great'). It has a capital M (all multiple units bigger than the base use capital letters with the exception of 'kilo-'). The Megameter is not used much in measuring since the term 'kilo' (from the Greek, *khilioi* meaning 'thousand') used in kilometre is sufficiently large to be used for all the distances on Earth! However, look out for the occasional use of enormous figures used on huge scale statistics, such as:

Situations may result in CO_2 emissions in the range 15–30 Pg C y^{-1} by the year 2100.
(Quotation from Royal Society Policy, document 10/01, *referring to carbon sinks mitigating global climate change)*

In this statement, the letter P (peta-) is used and means 10^{15}.

When considering measurement of length, the **metre** is taken as the standard unit and is subdivided into increasingly smaller units or multiplied into larger units. Some of the most commonly used terms are summarised in Table 3.

(You might like to know that the metre used to be defined as $\dfrac{1}{40\ 000\ 000}$ of the circumference of the Earth, through Paris: then it became the distance travelled by light in a vacuum during a time interval of 1/299 792 458 of a second! Well, that's scientific progress!)

Table 3 *Units of length*

Unit	Meaning	Factor	Symbol
Gigametre	Greek *gigas*, 'giant'	10^9 or 1 000 000 000 m	Gm
Megametre	Greek *megas*, 'great'	10^6 or 1 000 000 m	Mm
kilometre	Greek *khilioi*, 'thousand'	10^3 or 1000 m	km *(small k)*
metre		10^1 m	m
centimetre	Latin *cent*, 'hundred(th)'	10^{-2} m	cm
millimetre	Latin *mille*, 'thousand(th)'	10^{-3} or 0.001 m	mm
micrometre	Greek *mikros*, 'small'	10^{-6} or 0.000 001 m	μm
nanometre	Greek *nanos*, 'dwarf'	10^{-9} or 0.000 000 001 m	nm
picometre	Spanish *pico*, 'small bit'	10^{-12} or 0.000 000 000 001 m	pm

Test Yourself

Exercise 2.2.1

1 Put these into a series, with the largest first.

kilometre, nanometre, centimetre, metre, micrometre, picometre, millimetre

KEY FACT   *Note the spelling of the suffix '-metre'. We use the other ending '-meter' for items of equipment that measure, e.g. thermometer.*

2.3 Using scales and measuring biological objects

How do we use these units, and how do we calculate the size of very small things? This is often a cause for some concern amongst students, so try to imagine the **scale** or relative sizes of the following objects:

● a field

● a quadrat (used in ecology)

● a stamen of a flower

● a root hair cell

● a blood cell

● a ribosome (an organelle in a cell, where respiration takes place)

● a nucleotide (part of the DNA molecule).

If you know what all of these are, then it will be obvious that they form a descending scale of size from the field, which would be thought of in terms of metres (or even American miles if in Texas!), down to the dimensions of a molecule which would be measured in nanometres. (NB There are even smaller units that are useful for physicists who are interested in atoms and subatomic particles, but that is beyond the usual remit of biologists.)

Here are some examples of every-day biological things and their relative sizes:

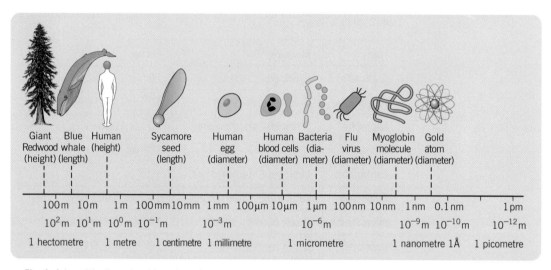

Fig. 1 *A logarithmic scale of length and some biological objects*

You may notice that the scale here is logarithmic (see p. 90 on logarithmic scales.) Each unit in *Fig. 1.* is 10 times bigger than the preceding one; used when the scales are very large – note particularly the use of the logarithmic scale in graphs.

One of the difficulties that biologists have to face is how best to express the size of an object. Let's take two extremes:

● the width of a cell membrane

● the height of an oak tree.

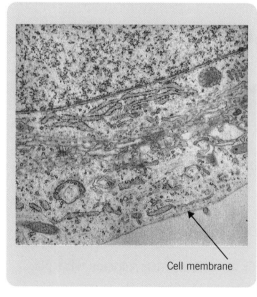

Fig. 2 *An electron micrograph of a cell*

Fig. 3 *An oak tree*

Clearly it would be inappropriate to express the size of the tree in nanometres or the membrane in metres! Imagine the figures, something like 0.000 000 003 m for the membrane or 500 000 000 nm for the tree!

What we need to do is to find the most appropriate unit to use in the circumstances.

Example
The average virus would measure about 0.1 μm diameter.
A typical eukaryotic cell could be about 50 μm diameter.
The length of a water flea (just visible to the naked eye) would be 1.0 mm.

Try a few conversions to become more confident in the use of this scale.

Test Yourself	Exercise 2.3.1

1 Simplify 0.000 23 mm

2 Convert 245 μm into millimetres (mm).

3 Express 2473 nm as micrometres (μm).

4 Which is bigger, 0.0023 mm or 23 μm?

KEY FACT

1 m = 1000 mm
1 mm = 1000 μm
1 μm = 1000 nm
1 nm = 1000 pm

2.4 Using units and symbols

One of the most useful skills in biology is to interpret measurements of tiny structures. Measuring the sizes from photographs or drawings enables us to understand the scale of the objects in real terms. It would not be possible to look down a microscope and appreciate the real size of a cell, for example, unless you could relate its size to a known measurement. If a student wrote down on a drawing that the cells they had shown were drawn using 'high power', there is little to explain how many of these cells, when put end to end, would cover a distance of 1 mm. On the other hand, the trained scientist, expressing the size of a cell as 'diameter 50 μm' should make more sense!

There are two ways in which you may need to calculate size in biology:

● to calculate from a drawing or actual photograph

● to use a microscope as a measuring tool.

In the case of a photograph, this will have been taken using either a standard light microscope or even an electron microscope. The latter is capable of magnifications up to hundreds of thousands of times. An electron microscope is used, not just because it can magnify more than a light microscope, but because it can distinguish between objects that are close together enabling the viewer to see, for example, a double membrane instead of just one line. This is known as the **resolving power** of the microscope. The light microscope is limited to discerning anything down to about 200 nm in size. However, the electron microscope can resolve objects down to 0.5 nm. Any two objects nearer to each other than the resolving power of the microscope will merge into one and will not be distinguished as two objects.

Some microscope photographs (micrographs) have a scale on them, while others just state the final magnification.

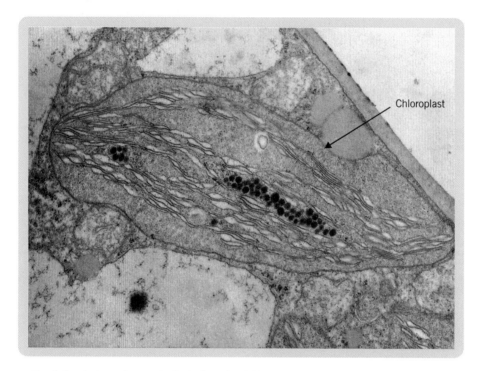

Fig. 4 *An electron micrograph of a leaf cell (×6000)*

This example shows how to calculate size from a scaled photograph:

Example
You are told that the magnification of this photograph of a plant cell is ×6000.

The chloroplast is labelled with the arrow. Calculate its maximum width and give your answer in micrometres (μm).

Method

1 First measure the width, which is about 4.5 cm. Remember to convert to millimetres (45 mm).

2 As the magnification is ×6000, you need to *divide* 45 mm by this to get the actual size in millimetres.

$\frac{45}{6000} = 0.0075$ mm

HINT

This number is not in μm. How many micrometres are there in 1 mm? There are 1000 so, to convert the 0.0075 mm to micrometers, you need to multiply by 1000. This is easily done my moving the decimal point three places to the right.

0.0075 × 1000 = 0.0̂0̂7̂5 = 7.5 μm

Example

You will often see in textbook illustrations (and for that matter in exam questions!) that a bar line has been drawn on the picture to show the scale. The line in *Fig. 5* is shown as 1 μm long. Using this line, it is possible to calculate the size of organelles in the picture and to work out the magnification of the photograph.

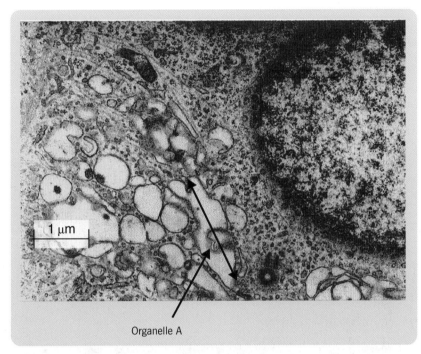

Organelle A

Fig. 5 *An electron micrograph of a cell*

Method

1 Measure the length of the line (scale bar) shown. It is 15 mm long, which is equivalent to 1 μm.

2 Measure the organelle labelled A on the photograph. You will find that it is 30 mm long (i.e. twice the size of the bar line). The size of this organelle A must be 2 μm long.

3 Now we will work out the magnification. We know that the line represents 1μm and is 15 mm long on the photograph and that there are 10^3 μm in 1 mm. Therefore, we need to multiply 15 by 10^3 to get the magnification.

   $15 \times 1000 = 15\ 000$ times

This is usually written '×15 000' and said 'times fifteen thousand'.

| Test Yourself | Exercise 2.4.1 |

1 In an electron micrograph of a cilium (magnification × 40 000), the diameter measured 12 mm. What was the actual diameter of the cilium?

2 A nucleus of a cell was shown as 2.3 μm, and measured 5 cm on the photograph. What was the magnification scale?

3 Which is bigger, a virus at 0.5 μm or a ribosome at 20 nm? Explain your reason.

HINT *Always convert to the same unit (millimetres here) before comparing sizes.*

Using a microscope to measure the size of cells

When you look down a microscope and see the enlarged view of a cell or organ, it is difficult to imagine how small it really is. You will know that the magnification of a microscope can be calculated by multiplying the magnification of the eyepiece by the magnification of the objective lens. Generally we use a ×10 eyepiece.

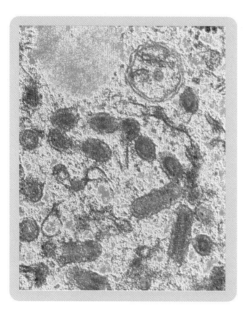

Fig. 6. *Rabies virus ×30 000*

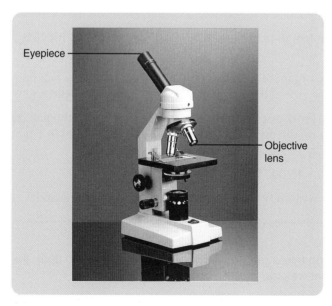

Fig. 7 *A microscope*

So if you were using the microscope with the ×10 objective and the ×10 eyepiece, the total magnification would be 10 × 10 = 100 times. (Usually the greatest magnification of microscope used at this level is ×400. What is the objective that would give this?)

However, this does not enable us to calculate the actual size of an object. To do this requires more calculations, but this is quite simple when you get the hang of it.

Task

Place a transparent plastic rule on the stage of the microscope. Use the lowest objective power. Look at the millimetre scale through the eyepiece. What do you see?

The field of view shows 7 units of the rule visible. That means that the distance across the field of view is 7 and 'a bit' units wide. (Don't forget it starts at 0!) Take the ruler away and then put an object underneath which may look like *Fig. 9*.

It is now possible to estimate the width of this object as 5 and 'a bit' units. If we could have left the rule in place and looked at the object at the same time, we could have used the scale to measure smaller sizes. (However, this is not possible because the ruler is quite thick and the object on a microscope slide would not be in the same focus.)

Well this is a little inaccurate isn't it? As scientists, we cannot say little things are 'four and a bit millimetres wide' can we? The element of error would be too great. To make it much more accurate we need something better than a plastic ruler and we also need some reference scale in the eyepiece too. Not all the objects we see down the microscope will conveniently fit the field of view!

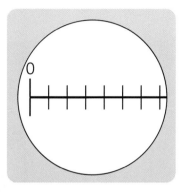

Fig. 8. *Magnified view of plastic ruler scale*

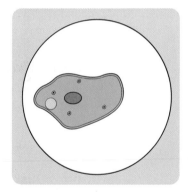

Fig. 9 *Small pond animal*

Using graticules and micrometers

No, don't panic! These are just fancy names for scales that are placed in the eyepiece and on the stage.

The eyepiece is unscrewed and the scale (graticule) dropped carefully in. As you can see the scale (which is finely etched on to either a glass or plastic disc) is a line, often about 1 cm long, divided into 10 units and also 100 smaller units (see *Fig. 10*).

The stage micrometer also has a scale, but this line is usually 1 mm long precisely and is accurately divided into 100 subdivisions – so each subdivision is 0.01 mm. It is mounted on a microscope slide with a very thin cover slip over it.

The first thing that needs to be done is the **calibration** of your microscope, because each one will differ slightly. Once you have done this it is not necessary to repeat it, provided you always use the same microscope to do your measuring.

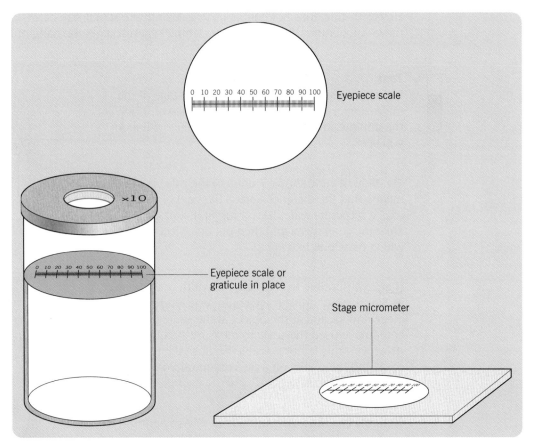

Fig. 10 *The graticule and stage micrometer*

Task

1 Place the micrometer slide on the stage of the microscope and focus, using the lowest objective lens (×4). Then change to the ×40 objective lens so that the magnification of the microscope is now ×400.

2 Move the slide about until both scales can be seen alongside each other, like this:

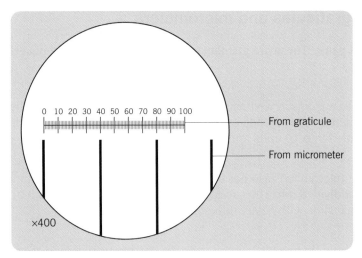

Fig. 11 *Parallel scales superimposed (1 micrometer unit = 40 eyepiece units)*

3 Now read off the scale and count the number of eyepiece divisions in 0.01 mm.

4 Calculate from the stage micrometer the size of the eyepiece divisions. (NB This scale in the eyepiece is only a reference scale, and is not to be taken as an actual length.)

5 Now you have calibrated the microscope for this magnification, repeat for the other magnifications (by changing to the other objective lenses ×4 and ×10).

6 You now know what the eyepiece divisions are equivalent to for each magnification of *that* microscope. Replace the stage micrometer with a slide of the cells to be measured. Line up the graticule scale in the eyepiece with the edge of the cell or organ, measure across and note the number of eyepiece divisions. Now you can calculate the actual size of the specimen.

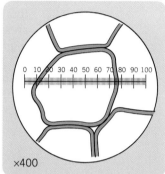

Fig. 12 *Size of cell – from 4–78 units*

Example
This is how to calculate the size of the cell in *Fig. 12*.

KEY FACT	The eyepiece graticule is usually 1 cm long and divided into 100 divisions or 'eyepiece units' (100 epu). One division of the stage micrometer = 0.01 mm.

Using one model of microscope it was found (*Fig. 11*) that 40 eyepiece units ≡ 1 stage micrometer division (i.e. 1 × 0.01 mm)

$$\text{So one eyepiece unit} = \frac{1 \times 0.01}{40} \text{ mm}$$

The cell (Fig. 12) is observed to be 74 eyepiece units (or divisions) across. So the width of the cell is 74 × 1 × 0.01 ÷ 40 mm or 74 × 1 × 0.01 ÷ 40 × 1000 μm.

The width of the cell we have measured is 18.5 μm.

Test Yourself

Exercise 2.4.2

1 When a certain type of microscope was calibrated with a ×40 objective lens and a ×10 eyepiece (i.e. magnification of ×400) it was found that 10 eyepiece graticule units were equivalent to 2 stage micrometer units.
A specimen when viewed under these conditions was observed to be 7 eyepiece units long. What was the length of the specimen in micrometres?

2 A specimen from a tropical pond (*Fig. 13*) was examined using the same microscope and magnification as in question 1. Estimate the width and length of the specimen in eyepiece units and convert these dimensions to micrometres.

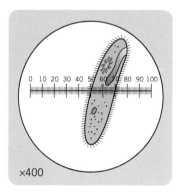

Fig. 13 *Paramecium* sp.

Counting cells using a microscope

In the previous examples, we have been using a microscope to measure small objects precisely. Another useful skill involves the microscope again, but this technique allows us to count cells, e.g. in a culture solution or blood corpuscles.

It is similar to a technique that you may have used in ecology where a quadrat marks out an area of a certain size, such as 1 m², so that an estimate or count can be made of the number of plants or animals in the enclosed area.

In the same way we can use this principle to count very small objects, such as cells, by using a microscope slide with a grid, like a quadrat, but scaled down so that it is only visible with a microscope.

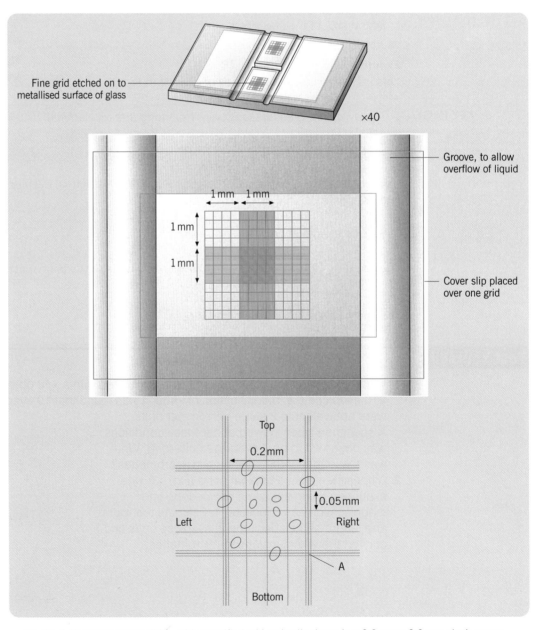

Fig. 14 A *Haemocytometer* slide – with magnified grid and cells shown in a 0.2 mm × 0.2 mm single square

To calculate the number of cells in a particular volume of fluid you need to know the dimensions of the grid and the depth of the liquid above the grid.

As you can see from *Fig. 14.*

If the depth = 0.1 mm
and the area of one single square (A) of the grid = 0.04 mm^2
The volume = 0.1 mm × 0.04 mm^2 = 0.004 mm^3

Example
Here is an example of the type of exam question that could be posed.
A sample of yeast cells was diluted by 10^{-5} and then placed on the haemocytometer slide grid used above. 5 squares were counted and the mean number of cells per square was 8.
How many cells were there in 1 mm^3 of the original, undiluted sample?

Method
1 There are 8 cells in a volume of 0.004 mm^3.

2 So, in 1 mm^3 there must be $\dfrac{8}{0.004}$ cells = 2000 cells.

3 The sample was diluted by 10^{-5}, so you must not forget to multiply by that dilution to get the true number of cells in the original sample.

   $2000 \times 10^5 = 200\ 000\ 000$ cells in 1 mm^3

(NB Expressed in scientific notation – see p. 27 that should be 2.0×10^8 cells in 1 mm^3 or 2.0×10^8 cells mm^{-3}.)

Test Yourself

Exercise 2.4.3

1 Fiona set up an experiment with some yeast cells in two different sugar solutions, to see which produced the greatest rate of growth.

She counted the cells in both cultures over several days using the haemocytometer slide:

Mean of number of cells per square of the grid, solution 1					
Day 1	Day 2	Day 5	Day 6	Day 7	Day 8
6.4	9.4	43.2	49.2	71.0	77.4

(In case you were wondering, Days 3 and 4 were at the weekend, and she wasn't that keen to get into the school!)

Her original data for one of the solutions are shown. Calculate the number of cells in 1 cm^3 of solution for days 1, 5 and 8.

All of the information so far in this chapter has been about one **base unit**, the unit of length – the metre – and its **sub-units**. The other base units in common use in biology courses are shown in Table 5, p. 29.

Shorthand symbols

However, before we leave this section on symbols, it may be worth reminding yourself of some others that you are likely to encounter in the mathematics of biology.

These are shorthand symbols and will be often found on graphs, in algebra and in textbook accounts. They save time and are a universal language amongst scientists and mathematicians. When you are taking notes, it is a good idea to use these and other abbreviations to save time and to allow you to concentrate on the technical details of the lesson.

Table 4 *Frequently used mathematical symbols*

$\propto$	is proportional to	$=$	is equal to
$\sim$	about the same size as	$\approx$	is approximately equal to
$\neq$	is not equal to	$\equiv$	is equivalent to
$>$	is greater than	$<$	is less than
$\gg$	is much greater than	$\ll$	is much less than
$\geqslant$	greater or equal to	$\pm$	plus or minus
Δx	a change in x	δx	a small change in x
$\therefore$	therefore	$\sqrt{x}$	the square root of x
$\bar{x}$	mean of the values of x	Σx	sum of all the values of x

2.5 Units

Changing units

The general public reacts with horror to any change in an established system. People felt comfortable with the old currency of farthings, pennies and shillings and with the old imperial system of measurement of lengths (12 in = 1 ft, 3 ft = 1 yd, 1760 yds = 1 mile). There were powerful complaints about moving to the 'more complex' decimal system (10 mm = 1 cm, 100 cm = 1 m, 1000 m = 1 km)! In a period of change some expensive mistakes can happen. You may recall the American 1999 Mars Orbiter. Some calculations were made in centimetres and others in inches and these were not converted. Instead of a Mars orbit, they had a Mars impact.

That is why the *Système International d'Unités (SI)* was introduced in 1960 to overcome the wide range of units that were in use at that time. Motorists in Britain are reluctant to see the end of road-signs showing miles and the American motorists cling avidly to both miles and gallons. However, scientists the world over use SI units and you too, in your science studies, must always use the SI system when measuring volume, distance, mass, etc.

In biology there are fewer examples of base units that need to be known than in the other sciences. They are summarised in the following table of measurements where the **base unit** is indicated in bold print.

Table 5 *SI units, base and derived, based on* Biological Nomenclature, *Institute of Biology, 2000*

Measurable feature	Unit name	Symbol of unit
Length	**metre** millimetre centimetre	**m** mm cm
Mass (NB Avoid confusion with weight.)	**kilogram** gram	**kg** g
Time	**second** minute hour	**s** min h
Area	square centimetre square metre hectare ($\equiv 1 \times 10^4$ m^2)	cm^2 m^2 ha
Volume	metre cubed (although cubic decimetres and cubic centimetres are also in common use)	m^3 dm^3 cm^3
Amount of substance	**mole**	**mol** (M)
Temperature	degrees Celsius (actually the SI unit is **kelvin**)	°C **K**
Pressure	**pascal** (newtons per metre squared (Pa = N m^{-2}) kilopascal	**Pa** kPa
Energy	**joule** (NB Calorie must not be used.)	**J**
Force (= mass × gravity)	**newton**	**N**

The base units of most interest to biologists are the metre, kilogram, second, (as well as the ampere, Kelvin, mole and candela). Sometimes it is more convenient to use multiples or sub-multiples of the base unit if the figures are very large or small. (See Table 3, p. 17.) For example the base unit for measuring time is the second, and for most biology experiments that would probably be fine. However, if you were measuring the effects of temperature on the growth of yeast in cultures over 5 days, the seconds would soon mount up! In this case the hour (h) is a much more useful unit. Note that a common mistake is to write '25 *grams*' instead of simply '25 g'.

Some rules for units

Capital letters for units or not?
The base unit and subdivisions all have initial lower-case letters, e.g. metre, millimetre, etc. The units bigger than the base unit usually have capital initial letters, e.g. Megametre (10^6 m) and Gigametre (10^9 m), but the exception is kilo- (10^3) as in kilometre and kilogram.

Capital letters for symbols or not?
The simple answer for this is – capital letters are not used for symbols unless the unit is named after a person. For example, the symbol for the unit for energy, the joule, is J, named after James Prescott Joule, and the symbol for the unit of force, the newton, is N after

Sir Isaac Newton. You will notice, however, that the name of the unit (joule, newton) is written in lower case. So when writing about a thousand joules or a kilojoule, the unit would be kJ (note that the lower-case letter k is used for the value, 'thousand'). By the way, don't fall into the mistake still made by some food labellers who mix together kilojoules and calories – the latter term is no longer used by scientists!

Fig. 15 *Well, they got most of it right!*

Look at some data put into a table by a student who was new to science. At first glance everything looks reasonable, but then some errors emerge.

Time (hrs)	Number of cells in culture solution		
	Temperature/10° centigrade	20 °C	30 °C
0	12	34	78
5	234	678	988
10	1,259	2,462	7,893

There are some simple illustrations here of points to be noted:

● The base unit for time is the second (s) but obviously the hour is more appropriate here.

● The unit for hour is not Hr or hr but simply **h** and it is better not to use the brackets but '/' as in 'time/h'.

● Temperature is often spoken of as centigrade. However, the correct unit is degrees **Celsius**.

● Spot any more points? How about the figures 1,259, 2,462 and 7,893? The comma is not correct, and it should be avoided altogether. The three numbers shown are conventionally written without a gap i.e. 1259, 2462 and 7893; but any numbers bigger than 9999 should be shown with gaps to separate the groups of three digits, such as 10 000 and 123 345 000.

Test Yourself

1 How would you show the following figures with the correct SI units?
(a) A temperature of 23 degrees.
(b) A mass of 34 kilograms
(c) The volume of a box, with all dimensions 4 cm.
(d) A pressure equivalent to 5 N m^{-2}

2 Correct the following, showing your answers as SI units:
(a) The energy of a peanut is expressed in J per gm.
(b) The density of water is 1 Kilogram per cubic metre.
(c) The solution contained 0.5 Moles/Litre of glucose.

3 State what is wrong with these sentences, in terms of SI terminology:
(a) After 24 hours the bacterial culture contained 2,342,000 organisms.
(b) The pressure in the autoclave reached 25 lb/in^2.
(c) Cornflakes contain 367 kcal per 100 g.

Derived units

Density is an example of a **derived unit** (or compound unit). It may be said as 'kilograms per cubic metre' and shown as **kg m^{-3}**. These units are composed of two or more base units (here, mass and length – the kilogram and the metre – but written with a space between kg and m). Biologists at this level also use the joule (symbol J) kg m^2 s^{-2} and the pascal (symbol Pa) kg m^{-1} s^{-2} (with spaces between the three symbols for the base units).

To 'per' or not to 'per'!

This is another case of a rule – but it is quite simple. It is acceptable when speaking to say that, for example, the amount of energy reaching the Earth can be measured in kilojoules *per* square metre. When we write it, however, we must express it as a compound derived SI unit, the word 'per' actually means 'divided by' or 'for every'. So, if you want to speak about energy reaching the Earth from the Sun, you could say it as *'kilojoules per square metre'* but to write it you would use the derived unit. You probably know that to divide something by 'm^2' is the same thing as multiplying it by 'm^{-2}' so you should write 'kJ m^{-2}'. If you are still uneasy about this, you may wish to consider the energy flow diagram given in *Fig. 17*.

Fig. 16 *purrrrr!*

This was taken from an examination question all about calculating the percentage of available solar energy trapped by plants. It is a common approach in ecological energetics questions. (However, you don't need to know about the ecology at this stage. The purpose here is to note the units and how they are expressed.) It states that the units are kJ m^{-2} year^{-1} which, when broken down, means, kilojoules per square metre per year.

Assume that 500 kJ solar energy arrived on 5 m^2 of the Earth's surface at a moment in time. We would say this as '500 kJ per 5 m^2'. However, if we wanted to express this not as 'per 5 m^2' but as 'per 1 m^2' what would we have to do?

500 kJ/5 m^2 can be simplified to 100 kJ m^{-2}

When we divide something by m² it is like 'x over y' or x^{-y}) so, it is the same thing as multiplying by m⁻². Whereas we would *say* '100 kJ per metre squared' we would *write* it as '100 kJ m⁻²'.

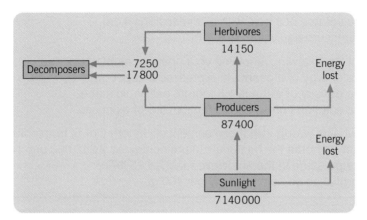

Fig. 17 *Annual energy flow, in kJ m⁻² year⁻¹, through an aquatic ecosystem (Data from AEB, Summer 1999, Paper 1)*

Fig. 18 *An example of a wetland ecosystem*

In many cases in examinations you will be asked to quote figures from tables or graphs, and the units may be given in the wording of the question.

Some other units less commonly found in biology

There are some units that only appear in one particular biological context. An example is in questions on osmosis and water relations. If you have already studied this, you will have come across the terms water potential, solute potential, etc. On p. 39 you will see an example using the symbol ψ for aspects of this relationship. **Water potential**, is a measurement of pressure, and the unit used is **pascal**, Pa, (or kilopascal, kPa.).

Another unit that you may come across is the measure of **frequency**, measured in **hertz**, Hz. This may crop up when discussing thresholds of hearing. Also the measurement of **force** may occur in some topics such as muscle action, then we would be using the unit **newton** (capital N remember, because it is named after Sir Isaac).

KEY FACT   *1 pascal is the same as 1 newton per square metre (Pa = Nm^{-2}).*

Finally, there is still confusion about weight and mass. For most biology (unless involving life on other planets or work in zero gravity), we should simply refer to mass all the time. It may be quite hard for us to drop the habit of referring to the 'weight of sucrose', or 'the plant lost 50 g in weight', etc. Just remember, always use the word 'mass'! Note also the terms 'biomass' and 'dry mass'.

Exam Questions

Q

Exam type questions to test understanding of chapter 2

1 *Fig. 19* shows the structure of a liver cell as seen using an electron microscope.

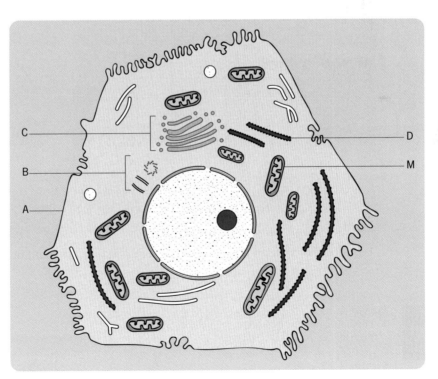

Fig. 19

The magnification of this diagram is ×12 000. Calculate the actual length of the mitochondrion labelled M, giving your answer in micrometres (μm). Show your working.

(Adapted from Edexcel, 1997, Paper B1)

2 A group of students sampled the animal life in a pond. They plotted their data as a pyramid of biomass as shown in *Fig. 20*. The biomass is expressed as milligrams per cubic metre.

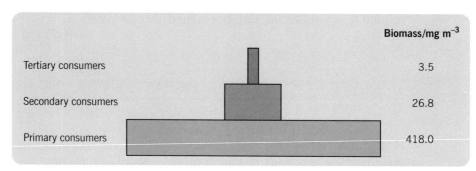

	Biomass/mg m^{-3}
Tertiary consumers	3.5
Secondary consumers	26.8
Primary consumers	418.0

Fig. 20 *Pyramid of biomass*

Calculate the percentage decrease in biomass between the primary consumers and the secondary consumers. Show your working (see percentage calculations, p. 5).

(Adapted from Edexcel, 1999, B2)

3 In an investigation, the absorptions of water and phosphate ions by the roots of a plant were recorded over a period of 24 h. The results are shown in *Fig. 21*.

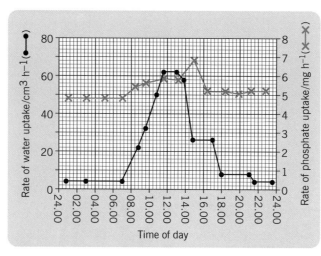

Fig. 21

Describe the changes in the rate of water uptake by the plant during the 24 hr period.

Exercise 2.2.1, *p.17*

1 kilometre, metre, centimetre, millimetre, micrometer, nanometer, picometer

Exercise 2.3.1, *p.19*

1 0.23 μm

2 0.245 mm

3 2.473 μm

4 23 μm

Exercise 2.4.1, *p.21*

1 $\dfrac{12}{40\,000} = 0.0003$ mm = 0.3 μm

2 50 mm × 2.3 × 1000 = 115 000; magnification × 115 000

3 20 nm = 20 000 μm, so the virus is much smaller!

Exercise 2.4.2, *p.25*

1 7 epu = $\dfrac{7 \times 2 \times 0.01}{10} \times 1000 = 14$ μm

2 20 epu × 90 epu; 40 μm × 180 μm

Exercise 2.4.3, *p. 27*

1 Day 1 = 1.6×10^6 cells, Day 5 = 1.08×10^7 cells;
   Day 8 = 1.94×10^7 cells
   (Hint – remember 1 cm^3 = 1000 mm^3)

Exercise 2.5.1, *p. 31*

1 (a) 23 °C (where C represents Celsius not
        centigrade)
   (b) 34 kg
   (c) 64 cm^3
   (d) 5 Pa

2 (a) $J\,g^{-1}$
   (b) 1 $kg\,m^{-3}$
   (c) 0.5 M glucose (no need to mention in what
        quantity, as this is assumed)

3 (a) 2 342 000 (no commas)
   (b) should be in Pa not lb/in^2
   (c) should be in kilojoules; (1560 kJ multiply by
        4.25)

Chapter 3

Algebra – x – the unknown!

After completing this chapter you should be able to:

- *understand the conversion of numbers to letters and back*
- *substitute into equations*
- *change the subject*
- *make an estimate of population size using mark-release-recapture method*
- *calculate the Hardy–Weinberg calculation (perhaps!).*

3.1 Equations and rules

Some students do not have very happy memories of working in algebra! They will be pleased that this is a fairly short chapter and, in fact, not many algebraic techniques are used in biology at this level. You will certainly know that we often use letters to represent numbers as in:

A snail crawls y cm along a straight line in z s. Calculate the speed of the movement. Let x denote the speed in centimetres per second (cm s^{-1}).

The letter x (or X) is frequently used for 'the unknown'. Other letters, either in the English or Greek (see p. 16) alphabet, can be used in equations. You can use any, provided that you state what each letter represents; take care, though, because some are in fairly standard use – as L and W for length and width in the next paragraph.

If you had to work out the surface area of a small pond, length 5 m and width 3 m, you would not find it too demanding to give an answer of 15 m^2. How did you arrive at that answer? Well, by multiplying the length by the width:

$$5 \text{ m} \times 3 \text{ m} = 15 \text{ m}^2$$

We can write that as a word equation:

Length $\times$ width = area

and to use just letters as symbols for the words we have:

$$L \times W = A$$

Already we have an algebraic equation, where the letters can be used to represent the numbers in a general situation. (The word 'equation' tells us that the value of the left-hand side of the equals sign (=) is the same as the value of the right-hand side.) Let's follow this aquatic example by finding the volume of a small aquarium. It is 80 cm long, 40 cm wide and filled to a depth of 35 cm. So we can write an algebraic equation:

$$V = L \times W \times D$$

By substituting the measurements for the aquarium we have:

$$80 \text{ cm} \times 40 \text{ cm} \times 35 \text{ cm} = 112\,000 \text{ cm}^3$$

Obviously, this is about as easy as it gets!

Substitution

Substituting in equations is straightforward. It is just what we have done: putting numbers instead of letters in the area and volume calculations, for example. The same equations could be used if you wanted to find the area of a forest or the volume of the nucleus of a cell (at least you could if the nucleus were a cube!). The numbers may be more complex and difficult to handle but the principle is the same. Substitute the correct number for a letter and it will help you to get the correct answer if you write in the units each time. You can see that in the earlier area calculation we were not only multiplying 5 by 3, we were also multiplying the units:

$$m \times m = m^2$$

Changing the subject

Still remaining with these 'easy number' examples, we need to think about **changing the subject** of the equation. The subject is the symbol on its own, on one side of the equals sign. Note that $L \times W = A$ is the same as $A = L \times W$. So in the second equation, the subject is A. Suppose that we are given the area of a lake (15 km^2) and can measure the length (5 km) but are unable to get to a position where we can measure the width. We could calculate the width by using the $A = L \times W$ equation and with simple numbers the answer is at once obvious. It is important to know the principles of changing the subject of an equation.

Fig. 1 *Lakeside view*

Two easy-to-remember rules are now about to emerge:

RULE 1	If you have an expression like $L \times W$ and you want to change it so that you only have W, then you must divide by L.
	So it is $\dfrac{\cancel{L} \times W}{\cancel{L}}$ which becomes W after cancelling.

If you do something to one side of the equals sign, the equation only gets back in balance if you do the same thing to the other side. So having divided one side by L, you must do the same with the other side:

$$\frac{\cancel{L} \times W}{\cancel{L}} = \frac{A}{L} \quad So, \ W = \frac{A}{L}$$

So, by substitution, we calculate that:

$$W = \frac{15 \text{ km}^2}{5 \text{ km}} = 3 \text{ km}$$

A friend said: 'I had to drive 235 miles and the air temperature was 86°!' Some people are so used to considering distances for journeys in miles and temperatures in degrees Fahrenheit, that they find it difficult to get into the habit of converting to kilometres and degrees Celsius. Algebra can help them. You can make a statement: 'Given a temperature in degrees Fahrenheit you convert it to degrees Celsius by first subtracting 32 and then dividing by 1.8.' Now if you wish to write that as a formula and work out what is the equivalent of 86 °F, you could proceed as follows:

Let F be the temperature in Fahrenheit and let C be the temperature in Celsius, then:

$$C = \frac{F - 32}{1.8}$$

By substituting we see that:

$$C = \frac{86 - 32}{1.8} = \frac{54}{1.8} = 30 \,°C$$

In this equation, the letters stand for **variables** because we can substitute *any* number that we need to calculate. The numbers in the equation are known as **constants**. When you multiply a variable by a constant, such as $b \times 5$, the constant is usually written first and the '$\times$' sign can be left out; so the above can be written as $5b$. Note also that if the constant is 1, we would not need to write it, for example, in $1c$ the constant is left out and it is correct to write just c.

Test Yourself

1 Given a distance in miles on road signs, you can convert to kilometres by multiplying by 1.6. Write this as an equation and work out the 235 miles in the statement above in kilometres.

2 Nutritionists use BMI (Body Mass Index) to work out whether a person is underweight, a healthy weight, overweight or obese. It is calculated by dividing the body mass in kilograms by the height squared in metres. Write the equation for BMI and work out the value of your own BMI.

Another convention that is generally accepted is that, unless we are explaining a method, we can leave out the multiplication sign in algebraic equations. So the earlier equation $V = L \times W \times D$ could be written with lower-case letters as $v = dlw$. Note that the order of the symbols has changed so that they are in alphabetical sequence – this is not important, all that really matters is that you can state what each letter stands for.)

Now you must deal with another rule:

RULE 3   *If you move one value from one side of the equals sign to the other side you must change the sign (+ becomes – and – becomes +). For example, if $a + b = c + d$ and you want to make a the subject of the equation, you can do so by moving b to the other side and changing its sign to minus:*

$$a = c + d - b$$

This type of equation, that has been rearranged so that we have only one value on one side and how to work it out on the other, is known as a **formula**.

Test Yourself
Exercise 3.1.2

In *Fig. 2*, cells A and B are shown: The water potential of A is –200 kPa and of B is –300 kPa. The formula for the calculation is shown below and questions about another two cells, 1 and 2. You don't need to know any biology to answer it. (Note that the Greek letter in this equation is ψ, which is pronounced 'ps-eye'.)

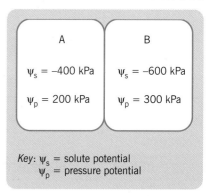

Key: ψ_s = solute potential
ψ_p = pressure potential

Fig. 2

Word equation: Water potential = solute potential + pressure potential

Equation with symbols: ψ = ψ_s + ψ_p

Cell 1 :   **–500 kPa**   =   **–600 kPa**   + ψ_p
Cell 2 :   **–500 kPa**   = ψ_s +   **200 kPa**

1 Calculate the value of the pressure potential (ψ_p) for cell 1.

2 Calculate the value of the solute potential (ψ_s) for cell 2.

Squares and square roots

There are two more rules which deal with squares and square roots in equations.

RULE 4   *If you have an equation with a value squared on one side, you can easily calculate its square root; but you must maintain the balance of the equation by finding the square root of the other side as well. For example, if $a = b^2$ then, in order to make b the subject, we must find the square root of b^2 (note that $\sqrt{b^2} = b$ in the same way as $\sqrt{3^2} = 3$:*

$$\text{So } \sqrt{a} = \sqrt{b^2} \text{ or } \sqrt{a} = b$$

> **RULE 5**   *If an equation includes a square root of a value, that value can be made the subject of the equation by squaring it and, of course, also squaring the other side of the equation. For example,*
>
> $$\sqrt{p} = rs \ \ so \ \ p = (rs)^2$$

Test Yourself

1 Use the equation: $a = \pi r^2$
   (a) Make r the subject of the equation.
   (b) Calculate the radius of a circular quadrat which would have an area of 154 cm².

HINT

Let radius = r and area = a. Use a value of $\frac{22}{7}$ for π (pi).

To divide a by $\frac{22}{7}$ is the same as multiplying it by $\frac{7}{22}$.

√ ((1 5 4 × 7 ÷ 2 2)) =

Of course, there is probably a key for π – in which case the key strokes are:

√ (1 5 4 ÷ shift π)) =

> **KEY FACT**   *Whatever you do to one side of the equation, you must do to the other.*

3.2 Equations and ecology

There is one important equation that ecologists use when working out the size of a population of motile animals. It is based on the use of ratios and proportions (see p. 51). It is said that a mathematician (before the days of census forms and electoral registers) worked out a method of estimating the population of Paris. He stood on a busy square and counted all those people who passed by. He recorded the number of those that were priests and the number of those not wearing clerical dress. The Archbishop knew the total number of priests in Paris, so the total population of the city could be calculated.

Let us look at an ecological study to explain this.

Field work

Pauline and Ali made a study of a rather old and crumbling moss-covered wall. They found a huge number of woodlice (*Oniscus asellus*). They collected 404 and marked each one on its underside with a small spot of harmless paint. They released their collection back to the wall habitat and returned 24 h later. This time they recorded:

Number of woodlice caught   = 297
Number of these with paint spot = 104

They estimated the total woodlouse population of the wall by using this equation to calculate the **Lincoln Index**:

Let N_1 be the number in the first sample captured and marked.
Let N_2 be the total in the second sample.
Let N_3 be the number of marked individuals in the second sample.
Then the total population will be:

$$\frac{N_1 \times N_2}{N_3}$$

By substituting, the total population of the wall is:

$$\frac{404 \times 297}{104} = 1154$$

This method is often known as the **mark-release-recapture technique**.

Test Yourself
Exercise 3.2.1

1 In a study of winter-feeding flocks, 36 blue tits visiting a bird table were trapped and before release each bird was marked by placing a small metal ring around one leg. The following day, 43 blue tits were trapped. Of these, 21 were already ringed. Estimate the size of the blue tit population visiting the bird table.

HINT *Write the equation first and then substitute.*

3.3 The Hardy–Weinberg equilibrium

This chapter ends with another biological topic that uses equations – but it only appears in some syllabuses. The **Hardy–Weinberg equilibrium** is a technique for working out the frequencies of alleles and genes in large and randomly mating populations. If a gene can exist in two forms (alleles), i.e. dominant and recessive, then the total number of alleles for that gene in the population would be 100% (obviously!). If you are told that the frequency, p, of the dominant allele, A, is 80%, then the frequency, q, of the recessive allele, a must be 20%.

So, $p\% + q\% = 100\%$. This can also be written as:

$p + q = 1$ [**Equation 1**]

The other equation that we need is worked out from the possible combinations of alleles to give genotypes. Thus

First allele and frequency		Second allele and frequency		Genotype and frequency	
A	p	A	p	AA	$p \times p = p^2$
A	p	a	q	Aa	$p \times q = pq$
a	q	A	p	Aa	$p \times q = pq$
a	q	a	q	aa	$q \times q = q^2$

If you have done the genetics of this (or else see p. 52) you will recognise these frequencies in the last column as the well-known 1 : 2 : 1 ratio. Using the same logic as earlier in this example, the frequencies of the genotypes in this population can be added to give 100% again, i.e. AA + Aa + aa = 100% (or 1.0) and from the table we can rewrite it as:

$$p^2 + 2pq + q^2 = 1 \quad \textbf{[Equation 2]}$$

Armed with equations 1 and 2 you can carry out some interesting calculations in population genetics.

Example

Here is one calculation that deals with the two-spot ladybird, which does such good work in clearing pests from rose bushes. It can either be red (rr) or black (BB or Bb) because of the dominant gene (B). In one year the frequency of the B allele was 0.4 (or 40%).

(a) State the frequency of the b allele.

(b) Calculate the expected frequencies of ladybirds with black bodies in the population and also ladybirds with the genotype Bb.

The only number that we have to start with is the frequency of the dominant allele B: we know that $p = 0.4$.

(a) Using equation 1, $p + q = 1$, q – the frequency of the recessive allele b – must be 0.6.

We can now use equation 2:

p^2	+	$2pq$	+	q^2	= 1
$(0.4)^2$	+	$2(0.6 \times 0.4)$	+	$(0.6)^2$	= 1
0.16	+	0.48	+	0.36	= 1
BB	+	Bb	+	bb	= 1

(b) So the frequency of the ladybirds with black bodies, i.e. BB + Bb, is 0.64 (64%). The frequency for those which have the genotype Bb is 0.48 (48%).

Test Yourself

Exercise 3.3.1

1 A population of lizards on an island has lizards with brown or yellow skin. The brown allele (B) is dominant to the yellow (b). The frequency of yellow lizards is 0.25. How many in a population of 2000 will be of the Bb genotype?

Exam Questions

Exam type questions to test understanding of Chapter 3

1 In estimating the size of a fish population, 70 fish were trapped, marked and released. A week later, a second sample was captured. Of these, 27 were found to be marked and 13 were not. Calculate the estimated size of the fish population. Show how you arrived at your answer.

(AQA, Human Biology, Summer 1998, Paper 1)

2 One evening, 54 garden snails were found in a small enclosed garden. They were each marked with a spot of white correction fluid. They were released. The following evening 35 were collected; only 7 of them were marked. Estimate the size of the snail population of this garden.

It is unlikely that other types of question that depend on the use of algebra, such as question 3, will be set in written examinations; but you should be able to manipulate equations and substitute numbers into formulae. The techniques could also prove essential in working out the results of your practical work and in the presentation and writing up of your research project.

3 Solve each of the following equations by finding x:

(a) $\dfrac{x}{15} = 3$

(b) $4x = 64$

(c) $x + 7 = 9$

(d) $x - 1 = 14$

(e) $\sqrt{x} = 11$

(f) $x^2 + 5 = 21$

4 A culture of flour beetles was maintained with 149 red beetles and 84 black beetles. The allele for red body colour, R, is dominant to the allele for black body. Use the Hardy–Weinberg equilibrium to calculate the expected number of Rr beetles in the population.

(Adapted from AQA, 2000, Paper 1)

Answers to Test Yourself Questions

Exercise 3.1.1, *p. 38*

1 The equation is $K = M \times 1.6$. The distance is 376 km.

2 $BMI = \dfrac{Mass}{Height^2}$ or $BMI = M/H^2$ (Only you will know the answer to this question – ideally it should be between 20 and 25 – a value of 27+ is regarded as obese.)

Exercise 3.1.2, *p. 39*

1 -500 kPa $+ 600$ kPa $= 100$ kPa

2 -500 kPa $- 200$ kPa $= -700$ kPa

Exercise 3.1.3, *p. 40*

1 (a) $\dfrac{a}{\pi} = r^2$ so $r = \sqrt{\dfrac{a}{\pi}}$ (b) 7 cm

Exercise 3.2.1, *p. 41*

1 Population $= \dfrac{36 \times 43}{21} = 74$

Exercise 3.3.1, *p. 42*

1 If $q^2 = 0.25$, then $q = \sqrt{0.25} = 0.5$. So $p = 0.5$ and $2pq = 2(0.5 \times 0.5) = 0.5$ or 50%. 50% of 2000 is 1000.

Chapter 4

Indices, powers of ten, scientific notation

After completing this chapter you should be able to:

- *use indices and powers of ten with confidence*
- *use scientific notation (standard form) correctly for biological structures*
- *carry out basic calculations with a scientific calculator*
- *understand positive and negative indices.*

These topics can often appear confusing in the early stages of study. So if you are not already familiar with them, you should gain confidence by starting with some basic skills and working through this chapter with care.

4.1 Indices

If you multiply 3 by 3 you have 'squared' the number 3 (think of a small rug 3 m by 3 m – it is 9 square metres in area). We can write this as

$3 \times 3 = 9$ or 3^2

Here the 3 is known as the **base** and the 2 is the **index** or **power** or **exponent** – all three terms can be used (the plural of index is **indices**).

Note that for 3^2 the index number 2, the sign for 'squared', is written or printed *above* the number to be squared (we say that the 2 is written or printed as a 'superscript').

Now, $3 \times 3 = 3^2$ (We say 'three squared'.)
$3 \times 3 \times 3 = 3^3$ (We say 'three cubed'.)
$3 \times 3 \times 3 \times 3 = 3^4$ (We say 'three to the power four'.)

Can you see the rule?

Fig. 1 *If each of the small cubes has sides measuring 1 unit, the area of one whole face is 3×3 or 3^2 units and the volume of the whole cube is $3 \times 3 \times 3$ or 3^3 units*

1 Calculate the following: (a) 2^3 (b) 2^5 (c) 4^3 (d) 17^2 (e) 9^5

4.2 Powers of ten

It follows from the introduction above that using indices could simplify matters. Just imagine reading a magazine article:

> It is likely that all matter originated in the big bang that happened 13–20 000 000 000 years ago and that the Earth was formed 4 600 000 000 years ago. The first living organisms were around some 4 000 000 000 years ago. It is important to realise that the mean distance from the Earth to the Sun is 149 500 000 km.

Did you get fed up with all of those zeros? There is a much easier way to write such numbers and it is to use powers to the base of ten.

Obviously, you can work out 10×10 and realise that this can also be written 10^2.

So you will understand that

$10 \times 10 = 10^2$ (hundred)
$10 \times 10 \times 10 = 10^3$ (thousand)
$10 \times 10 \times 10 \times 10 = 10^4$ (ten thousand)
$10 \times 10 \times 10 \times 10 \times 10 = 10^5$ (hundred thousand)
$10 \times 10 \times 10 \times 10 \times 10 \times 10 = 10^6$ (million)

Again, the rule is clear. This device allows you to write numbers in a different way.

Instead of eight thousand you could write 8×10^3. How could you write that the population of Bristol is six hundred thousand and that the UK population is fifty million? Try it!

What about really big numbers? The adult human body has about fifty million million cells. So how do we write that? Well knowing that a million is 10^6, it would be:

$50 \times 10^6 \times 10^6$

The above would be correct but it could be simplified even more to 50×10^{12} (but note p. 48).

The cell number example shows us a useful rule:

KEY FACT   *We can multiply together two big numbers expressed as powers of ten just by adding the indices.*

So, $10^5 \times 10^3 = 10^8$ and $10^{25} \times 10^{33} = 10^{58}$, but this number is astronomical!

I Calculate: (a) $10^4 \times 10^3$ (b) $10^6 \times 10^{15}$ (c) $10^7 \times 10^3 \times 10^{12}$

Easy, isn't it? Can you think what the rule may be about dividing, say 10^4 by 10^3?

Scientists use the metric decimal system and it is important to get the units right.

Example
Given that one metre (1 m) is one thousand millimetres (10^3 mm) and that I am 1.7 m tall, what is that in millimetres?

Example
An elephant has a mass of 1.2 tonne. State the mass of the elephant in kilograms and grams (given that 1 tonne = 10^3 kg and that 1 kg = 10^3 g).

Fig. 2 An African elephant

KEY FACT   *To multiply such complicated numbers the rule is just add together the indices of the powers of ten and multiply the other two numbers. Thus:*

$$(3 \times 10^5) \times (2 \times 10^3) = 6 \times 10^{(5+3)} = 6 \times 10^8$$

Test Yourself   Exercise 4.2.2

1 Calculate these: (a) $(1.5 \times 10^5) \times (6 \times 10^8)$ (b) $4.3 \times 10^7 \times 1.2 \times 10^6$

HINT   **It may be better to write the second calculation with brackets first.**

In the examples above we are mixing powers of ten with numbers (1.7×10^3 and 1.2×10^6). The rule is simple:

HINT

As soon as you are familiar with these principles find out how to do similar calculations with your scientific calculator. Calculators vary; now we are going to find out how to enter 800 000. Of course it will be 8×10^5 – but how do you do that?

Try using the $\boxed{y^x}$ key (or $\boxed{x^y}$ key). Key in: $\boxed{10}\ \boxed{y^x}\ \boxed{5}\ \boxed{=}$ What number is displayed?

Your calculator may use an $\boxed{EXP}$ key or an $\boxed{EE}$ key to do this operation.

Try: $\boxed{8}\ \boxed{\times}\ \boxed{EXP}\ \boxed{5}\ \boxed{=}$, (or $\boxed{8}\ \boxed{\times}\ \boxed{EE}\ \boxed{5}\ \boxed{=}$), also $\boxed{8}\ \boxed{EE}\ \boxed{5}\ \boxed{=}$, $\boxed{8}\ \boxed{\times}\ \boxed{10}\ \boxed{EE}\ \boxed{5}\ \boxed{=}$

When you have sorted this out using the instruction booklet, try some of the earlier calculations and write out some of your own. Some people who feel a bit unsure of the technique should first deal with the powers of ten 'by hand' and then multiply the remaining numbers.

Negative indices

All of the above deal with handling numbers greater than 1. Look at *Fig. 3*.

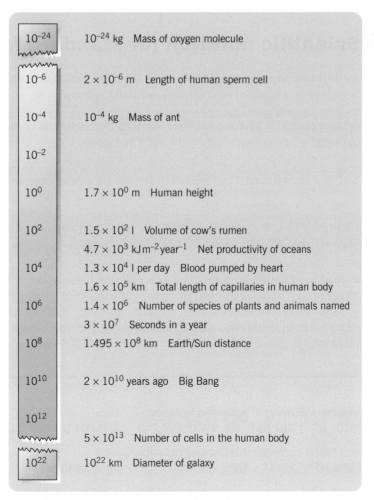

10^{-24}	10^{-24} kg Mass of oxygen molecule
10^{-6}	2×10^{-6} m Length of human sperm cell
10^{-4}	10^{-4} kg Mass of ant
10^{-2}	
10^{0}	1.7×10^{0} m Human height
10^{2}	1.5×10^{2} l Volume of cow's rumen
	4.7×10^{3} kJ m^{-2} year^{-1} Net productivity of oceans
10^{4}	1.3×10^{4} l per day Blood pumped by heart
	1.6×10^{5} km Total length of capillaries in human body
10^{6}	1.4×10^{6} Number of species of plants and animals named
	3×10^{7} Seconds in a year
10^{8}	1.495×10^{8} km Earth/Sun distance
10^{10}	2×10^{10} years ago Big Bang
10^{12}	
	5×10^{13} Number of cells in the human body
10^{22}	10^{22} km Diameter of galaxy

Fig. 3 *Some numbers on the powers of ten ruler.*

As biologists you will often have to manipulate some small numbers as well. Imagine that a human sperm cell is 2.5 thousandths of a millimetre across. Powers of ten can handle this – but to complicate matters we have to use **negative indices**. Table 1 should help to clarify this concept.

Table 1 *Powers of ten*

10^3	10^2	10^1	10^0	10^{-1}	10^{-2}	10^{-3}	10^{-4}
1000	100	10	1	0.1	0.01	0.001	0.0001
				$\frac{1}{10}$	$\frac{1}{100}$	$\frac{1}{1000}$	$\frac{1}{10\,000}$
				$\frac{1}{10^1}$	$\frac{1}{10^2}$	$\frac{1}{10^3}$	$\frac{1}{10^4}$

You can see from this that 'ten to the power minus two' 10^{-2} is the same as one hundredth or 'one divided by ten squared'.

KEY FACT *Any base to the power 0 (e.g. 2^0 or 3^0) is equal to 1. So $10^0 = 1$.*

By the way, the human sperm mentioned above is 2.5×10^{-6} m or 2.5 millionth of a metre.

4.3 Scientific notation (or standard form)

Make an effort to understand this topic – it is the source of some of the most common errors in the early days of a study of biology. It is, actually, quite simple.

When we use powers of ten for a number, such as the earlier 50 million million (see p. 45), we should write it *not* as previously (50×10^{12}) but as 5.0×10^{13}.

KEY FACT *Only use numbers above 1 but below 10 before the power of ten.*

KEY FACT *There can only be one non-zero number before the decimal point, e.g. 115 must be shown as 1.15×10^2 and 0.000 32 must be 3.2×10^{-4}.*

When we read in scientific notation that a cauliflower has a mass of 1.52×10^3 g we can rewrite that number in decimal form by moving the decimal point three places to the right: 1520.0 g. As you will now realise, the reverse is also true. So if you are told that a prokaryotic cell is 5.0×10^{-6} m in length you would have to move the decimal point six places to the left: 0.000 005 m (see p. 20), 5×10^{-6} is written with the SI unit 5 μm.

Test Yourself Exercise 4.3.1

1 Express the following in scientific notation:
 (a) 8970 (b) 1 467 851 (c) 3 461 897 213 (d) 0.001 01 (e) 0.046 (f) 0.000 000 712

2 Express these *correctly* in scientific notation:
 (a) 234×10^2 (b) 11×10^6 (c) 481×10^{-6} (d) 0.0043×10^{-3} (e) $0.012\,01 \times 10^4$

3 Express the following as ordinary decimal numbers (no powers of ten):
 (a) 4.3×10^{-1} (b) 8.761×10^{-4} (c) 6.2×10^4 (d) 0.0031×10^{-2} (e) 0.0121×10^2

And what next?

Find out how to manipulate negative indices on *your* calculator.

Try to multiply 10^3 by 10^{-2}. It is obviously $10^{(3-2)} = 10^1$.

Now you know that you can multiply using positive and negative indices with powers of ten, just by adding together the indices, but what about dividing? This is how you divide 10^5 by 10^3:

$$10^5 \div 10^3 = 10^5 \times 10^{-3} = 10^{(5-3)} = 10^2$$

You have to bring the number below the line (the denominator e.g. 10^3) above the line, so change the sign and then add the indices: $(5 + -3) = (5 - 3)$ as shown above.

KEY FACT | *To divide using scientific notation and more complex numbers you do the same thing with the powers of ten and deal with the numbers separately:*

$$(4.2 \times 10^8) \div (2.1 \times 10^5) = (4.2 \div 2.1) \times 10^{(8-5)} = 2 \times 10^3$$

HINT | *Using my calculator I would do that by using the following keys:*

(4 . 2 × 1 0 EXP 8) ÷ (2 . 1 × 1 0 EXP 5) =

Test Yourself

Exercise 4.3.2

1 Calculate the following:
   (a) $(7.5 \times 10^4) \div (2.5 \times 10^2)$
   (b) $(5.2 \times 10^{11}) \div (1.7 \times 10^{10})$
   (c) $(9.2 \times 10^3) \div (4.0 \times 10^2) \times 6.1$

Now check back to Chapter 2 and think about using numbers with units, not just 1.7 m for length and 1 g for mass – but kilometres per hour for speed (or indeed metres per second). Remember that you must *never* use the '/'. It is not 30 km/h but 30 km h^{-1}.

HINT | *Using Microsoft Word, 10^2 can be typed as: 10 (ctrl and shift keys together with +) followed by 2 for the index and then you can cancel the superscript by again typing together (ctrl/shift/+).*

Exam Questions | **Exam type questions to test understanding of chapter 4**

1 Artificial insemination is the transfer of a semen sample to the female reproductive tract without the use of a male partner, the male acting simply as a donor of sperm. The table shows some of the characteristics of semen from cattle.

Ejaculate volume/cm^3	5
Sperm concentration/number per cm^3	1.1×10^9
Percentage of motile sperm	70

   (a) Ten million motile sperm are required per cow for artificial insemination (AI). Use the information in the table to calculate how many cows can be artificially inseminated per ejaculate. Show your workings.
   (b) Suggest two advantages of using artificial insemination in cattle.

(AQA, BY02, June 1999)

2 A student was looking at the epidermal cells in the coleoptile of a barley seedling. The whole coleoptile was first measured after 12 days of germination and found to be 19 mm long. Using the graticule of a microscope, she measured several cells and recorded their length as 120 μ. How many cells form a row the length of the coleoptile? In a second experiment, where the plants had been given extra nitrate, the mean length of the coleoptiles was 26 mm; the number of cells in a row was the same as in the first experiment. Calculate the epidermal cell length and show all of your workings.

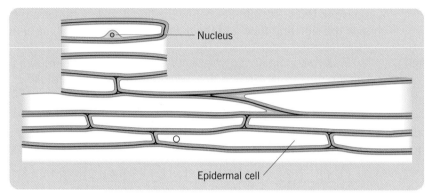

Fig. 4 *Epidermal cells*

Exercise 4.1.1, *p. 45*
1 (a) 8 (b) 32 (c) 64 (d) 289 (e) 59 049

Exercise 4.2.1, *p. 45*
2 (a) 10^7 (b) 10^{21} (c) 10^{22}

Exercise 4.2.2, *p. 46*
1 (a) 9.0×10^{13} (b) 5.16×10^{13}

Exercise 4.3.1, *p. 48*
1 (a) 8.970×10^3 (b) $1.467\,851 \times 10^6$
  (c) $3.461\,897\,213 \times 10^9$ (d) 1.01×10^{-3}
  (e) 4.6×10^{-2} (f) 7.12×10^{-7}

2 (a) 2.34×10^4 (b) 1.1×10^7 (c) 4.81×10^{-4}
  (c) 4.3×10^{-6} (d) 1.201×10^2
3 (a) 0.43 (b) 0.000 876 1 (c) 62 000
  (d) 0.000 031 (e) 1.21

Exercise 4.3.2, *p. 49*
1 (a) 3.0×10^2 (or 300) (b) 3.06×10 (or 30.6)
  (c) 23×6.1 (or 140.3)

Chapter 5

Ratios and getting things in proportion

After completing this chapter you should:

- *understand the meaning of the term ratio*
- *be able to calculate simple numeric ratios*
- *be able to translate data from breeding experiments to 3 : 1 ratio and 9 : 3 : 3 : 1 ratio*
- *understand surface area/volume ratios*

5.1 Ratios

How much cooking do you do? Some people tend to improvise in the kitchen. The food may turn out OK most of the time; however, since they don't follow a recipe, they may never seem to be able to repeat it exactly the second time round. If you look at an old recipe you find some interesting measurements such as cup, spoonful, handful and pinch! These are not really very accurate – but is it really necessary to weigh out accurately each ingredient? As long as the measure used is standard, i.e. if the cup is the same each time, then it is the **proportion** of the ingredients that is important.

There is an advantage in having proportions stated in recipe books, because you can adjust the quantities of the ingredients according to how much you want to make. For example, someone may want to enter the *Guinness Book of Records* for baking the biggest cake in the world. In that case, any recipe can be adjusted so that the proportions remain the same; the quantities could be changed from grams to kilograms or even to tonnes of flour, eggs etc. Proportion can be stated as a fraction – for instance a fruit pie could consist of $\frac{2}{5}$ pastry and the rest fruit; so the proportion of the pie that is fruit is $\frac{3}{5}$.

Although the idea of proportion is a useful concept, in biology you will often use the **ratio** of one quantity to another instead. So, you could say that the ratio indicating the relationship of one quantity to another in the fruit pie is: 2 parts of pastry to 3 parts of fruit. Ratios help to establish patterns without getting us bogged down with the fine details. Here is an example you may find useful – even if not in biology!

Example

If you want to mix cement there are different ratios of ingredients according to what mix you want to make. Cement for bricklaying requires 1 part cement to 3 parts sand to 1 part lime; whereas if you were making the foundations for a path, you would use 1 part cement to 5 parts coarse gravel! (Who says education is pointless?)
Another way of writing these figures would be:

Use of mix	Volume of cement	Volume of sand or gravel	Volume of lime
Bricklaying	1	3	1
Path foundations	1	5	0

The first ratio can be read as '1 part to three parts to 1 part'. Another, shorter way of showing this is:

Cement : sand : lime = 1 : 3 : 1
This is read as 1 is to 3 is to 1.

You can see that there are no real measurements here; the unit could in fact be a shovelful or a bucket, it really doesn't matter, provided you use the same measure each time. When we write ratios we use the colon (:) and dispense with the units.

5.2 The use of ratios in genetics

Probably the most 'famous' use of ratios in biology must be as a result of the work of the 'father of genetics', Mendel. Johann Mendel was born in 1822, in Moravia, the son of a farmer. He took the name Gregor when he became a Monk. Frustrated by family poverty he studied and qualified in theology, and started teaching mathematics and Greek at a grammar school. He then moved to Brno to teach agriculture and natural history. However he was unsuccessful in his examinations to become a qualified teacher and consequently took on the role of monastery kitchen gardener and beekeeper – where he could continue his plant breeding investigations. It is worth noting that while he was trying to become a teacher, his examiner wrote 'he lacks insight and the requisite clarity of knowledge' and failed him!.

Around 1856, Mendel started his experiment on garden plants. The most successful plants turned out to be the garden pea (*Pisum sativum*) You will know the story: he initially observed that there were recognisable differences between individual pea plants, such as stature of plant. He wrote,

In experiments with this character, in order to be able to discriminate with certainty, the long axis of six to seven feet was always crossed with the short one of $\frac{3}{4}$ foot to $1\frac{1}{2}$ feet.

(These were his 'tall' and 'short' plants.)
We will not go into the details here, as most textbooks cover this well, but it is interesting to see what he actually wrote about the results of these experimental crosses.

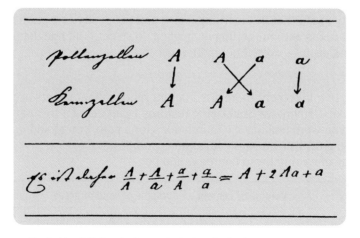

Fig. 1 *Mendel's handwritten calculations 1 : 2 : 1*

He also wrote:

> Out of 1064 plants, in 787 cases the stem was long, and in 277 the stem was short. Hence the mutual ratio of 2.84 to 1 ... If now the results of the whole of the experiments be brought together, there is to be found, as between the number of forms with the dominant and recessive characters, an average ratio of 2.98 to 1 or 3 to 1.

He had spotted a simple numeric ratio; what was so remarkable about this work was that it developed into what we now accept as modern genetics. The algebraic symbolism, the statistics, the clarity of explanation and the choice of seven contrasting characters in the pea to study cannot have been due to luck alone. We accept that he was a genius – but perhaps some of his published ratios may have been just a bit too perfect!

However, to get back to the idea of ratios. After thousands of garden plants had been interbred he eventually published his findings and laid the foundations for the modern science of genetics. In this publication he included the two famous ratios of:

$$3:1$$
$$\text{and } 9:3:3:1$$

We should now look in more detail at Mendel's results and see how a ratio can be established.

In Table 1 below, some of the actual results that he published for the inheritance of three of the seven pairs of contrasting characters are shown. Parents of the differing types were crossed and the appearance of the individual plants in the offspring was recorded.

Table 1 *Some of Mendel's results*

Characters crossed	Parental appearance	Offspring appearance
Length of stem	tall × dwarf	787 tall : 277 dwarf
Shape of seed	round × wrinkled	5474 round : 1840 wrinkled
Colour of seed	yellow × green	6022 yellow : 2001 green

How do we calculate the ratios of these figures? Does it matter that each set comprises different totals? How precise does a ratio need to be to be useful?

Example
Let's take the length of stem.

The figures show 787 tall to 277 dwarf. Divide both 787 and 277 by the smaller number. What do you get?

$\frac{787}{277}$ and $\frac{277}{277}$ or 2.84 and 1

So the ratio of the two numbers to each other is 2.84 to 1 or 2.84 : 1.

We can simplify this by rounding up to the nearest whole number. So 2.84 becomes 3 and the ratio can be expressed as:

$$3:1$$

If you are unhappy about the accuracy of rounding up or down see p. 135. For the purpose of comparison using ratios is quite an acceptable process, but it may need some statistics to find out how confident we are about our assumptions. More on that later.

Exercise 5.2.1

1 Calculate the ratios for the other figures in Table 1, they are again:
(a) 5474 to 1840   6022 to 2001

2 Mendel looked also at the position of the flowers. The results from a cross between axial flowers and terminal flowers were 651 axial and 207 terminal. What ratio is this?

3 In a similar cross involving colour of flower (red and white), 929 plants were produced. A ratio of 3 : 1 was discovered. How many red and how many white-flowered plants were there?

Fig. 2 *Garden pea flowers*

These are all examples of **Monohybrid inheritance** and, as Mendel found, the ratio normally tends to come out the same – three to one. (He was lucky in his choice of garden peas as they are relatively uncomplicated plants! However, he did cross many other species as well.)

This is only the start of the story – Mendel worked with other ratios. We can express ratios between any number of quantities. (Remember the example? There were *three* ingredients: cement, sand and lime, in a ratio of 1 : 3 : 1.)

> **KEY FACT**   *Ratios apply to any number of relative quantities. The only rule is that you divide each number in the set by the smallest one.*

The other ratio that Mendel gave to genetics involves a cross where *two* characteristics are considered together.

The dihybrid ratio in genetics

At this stage in your studies you may not have studied genetics – but you only need to follow the logic. So in this example, we take two contrasting characteristics – seed colour and seed shape.

In one batch of 556 plants that Mendel collected, he found the following seed characters present:

Round and yellow   = 315
Wrinkled and yellow = 101
Round and green   = 108
Wrinkled and green = 32

Fig. 3 *Pea seeds – round and wrinkled varieties*

The proportions of the four categories could be expressed as:

315 to 101 to 108 to 32 or even simplified down to 3.15 to 1.01 to 1.08 to 0.32

That would be a very clumsy ratio. We need to state a ratio where the smallest number is 1 and (using the rule) we can get this by dividing all four numbers by the smallest number, i.e. 32. What does this give?

$\frac{32}{32}$ = 1

$\frac{108}{32}$ = 3.38

$\frac{101}{32}$ = 3.16

$\frac{315}{32}$ = 9.84

Although not a perfect ratio, this could be seen as approximating to 9 : 3 : 3 : 1.

From this sort of information, it is possible to make predictions about the population in general.

Example

If we look back at the tall : dwarf = 3 : 1 ratio on p. 53 we can estimate that out of the whole population it would be expected that $\frac{3}{4}$ of the offspring would be tall and $\frac{1}{4}$ dwarf.

It is a bit more complex when we calculate the dihybrid population. You can see from the results above that:

The number of round seeded plants is 315 + 108 = 423
The number of wrinkled seeded plants is 101 + 32 = 133

In other words, there is still a 3 : 1 ratio (or a $\frac{3}{4}$ proportion of round seeded plants) which is unchanged even in the presence of the colour characteristic. Similarly

Yellow (315 + 101) : green (108 + 32) = 3 : 1

We can sum this up by showing the proportions expected for each characteristic as:

Round $\quad = \frac{3}{4}$

Yellow $\quad = \frac{3}{4}$

Wrinkled $= \frac{1}{4}$

Green $\quad = \frac{1}{4}$

The **probability** of combining these characteristics in a population is calculated by multiplying the two probabilities, thus:

Round and tall $\quad = \frac{3}{4} \times \frac{3}{4} = \frac{9}{16}$

Round and dwarf $\quad = \frac{3}{4} \times \frac{1}{4} = \frac{3}{16}$

Wrinkled and tall $\quad = \frac{1}{4} \times \frac{3}{4} = \frac{3}{16}$

Wrinkled and dwarf $= \frac{1}{4} \times \frac{1}{4} = \frac{1}{16}$

Now you can see clearly how Mendel worked using his knowledge of mathematics to derive the 9 : 3 : 3 : 1 ratio.

(Note that Mendel wrote that the characteristics were produced by 'factors' – he knew nothing about DNA – now we use the term **gene** and call the contrasting ones **alleles**, we understand the human **genome** and can work out family pedigrees.)

Test Yourself

Exercise 5.2.2

1 In *Drosophila melanogaster* (the fruit fly) crossing males and females with particular characteristics gave the following results:

Normal wing, white eyes $\quad = 35$
Normal wing, red eyes $\quad = 16$
Miniature wing, white eyes $= 17$
Miniature wing, red eyes $\quad = 37$

What is the ratio of these figures?

5.3 Ratios in other biological examples

Ratios are often used when the main point to get over is the relationship, as in this data collected in the laboratory. The precise figures can be calculated from the ratio.

HINT

Ratios give us the relative values of two or more quantities. Ratios tell us how much larger or smaller one quantity is than the others. There is no need to use any units in ratios, as they are simply expressions of relationships to each other.

Data from lab. work

Test Yourself

1 Kitty made a slide of the cells at the root tip of an onion and found cells in different stages of the cell cycle.

Stage of mitosis	Number of cells found
Interphase	140
Prophase	70
Metaphase	10
Anaphase	5
Telophase	15

Calculate the ratio of the number of cells in each stage.

Example
Another example involves quantities of gases in a mixture. If the ratio of nitrogen to oxygen to carbon dioxide in a sample of air is 1560 : 400 : 1, what would be the amount of oxygen in a 10 m^3 sample?

Method
Let's go through the process of working this out.

1 We are told that the ratio is 1560 : 400 : 1. This is not an easy set of numbers and could be simplified. First add up all the numbers, the answer is 1961.

2 Therefore out of a total of 1961 units of air, there are $\frac{1560}{1961}$ units of nitrogen.

3 $\frac{1560}{1961} = 0.80$

4 To express this as a percentage: $0.80 \times 100\% = 80\%$

5 If you do the same for the other gases, oxygen comes out as $\frac{400}{1961} = 0.20$, or 20%; and carbon dioxide comes out as $\frac{1}{1961} = 0.0005$, or 0.05%.

6 So, if oxygen comprises 20% of the sample, and the sample is 10 m^3 in total, the volume of gas that is oxygen would be:

$\frac{20}{100} \times 10$ m^3 = 2 m^3

(N.B. When rounding, percentages can sometimes add up to more than 100%.)

1 Refer back to Exercise 5.3.1, on the previous page.
  (a) Calculate the percentage of time spent in each phase.
  (b) Now, assume that a complete cell cycle takes 24 h and work out how much time is spent in interphase and in each of the stages of mitosis.

2 Soil is made up of a number of components, with clay particles, sand particles and humus being the main ones.

Look at the data for three different samples of soil.

Sample	Clay particles %	Sand particles %	Humus %
Garden in London	85	10	5
From moorland	2	4	94
Near river in Bristol	12	65	23

  (a) What is the ratio of soil components for the London garden?
  (b) In the soil from Bristol, what would be the mass of clay in a sample of 15 kg?

Surface area : volume ratios

One of the interesting features of all living things is that in order to obtain oxygen for respiration, they must absorb the gas through an exchange surface. This process is known as **diffusion**. So why is it that some animals have a simple body surface but others need to have very complicated gills or lungs that allow gases to pass through?

Partly it depends on where they live, either in water or in air; partly it is explained by the ratio of their surface area to their volume.

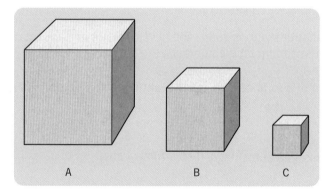

Fig. 4

The largest cube in *Fig. 4* has sides that are 3 units long, the next has sides of 2 units and the smallest has sides of 1 unit. What is the **surface area** of each cube?

  A is $9 \times 6 = 54$ square units
  B is $4 \times 6 = 24$ square units
  C is $1 \times 6 = 6$ square units

We could say that there is a relationship here, as the linear measurements increase by a ratio of $1 : 2 : 3$. The area ratio increases by $1 : 4 : 9$.

This is a particularly important feature of all living things that need to obtain oxygen through their outside covering. If we consider the *volumes* of these same cubes, and work out the ratios of volumes, we get:

A is $3 \times 3 \times 3 = 27$ cubic units
B is $2 \times 2 \times 2 = 8$ cubic units
C is $1 \times 1 \times 1 = 1$ cubic unit

So, whereas the areas had only increased to the ratio $1 : 4 : 9$, there is a much greater increase proportionally in the volumes of $1 : 8 : 27$. You can see that when considering the 3^3 structure, the volume needing oxygen (and getting rid of wastes) increased 27 times but the surface area only increased 9 times. This increase in surface area may be inadequate for big organisms; a larger exchange surface may be necessary, such as gills or lungs.

The same reasoning is true for surface area to volume ratios of spherical objects, where the volume is calculated by the formula $\frac{4}{3}\pi r^3$ or cylindrical bodies where it is $\pi r^2 h$.

Scales and measuring

A page may show a diagram of a very small water creature and on the next page the same space could be used for a picture of an elephant. The one is drawn larger than life size and the other, much reduced to fit the page.

Ratios are at work here. To work out the real size of these objects, we need to be aware of the **scale factor**. This is the ratio of the size of the original to the size of the drawing or photograph.

Example
Measure the length of the invertebrate in *Fig. 5*.

Your measurement should be 4.4 cm including the legs.

Now measure the scale line marked '1 cm'.

You should find that this is in fact 2 cm long. It has been scaled up because the animal here has been drawn twice the normal size.

So all you have to do is to divide the measured length of the drawing by the scale factor, the **magnification** (×2), and obtain the real length.

So the animal is $\frac{4.4}{2} = 2.2$ cm long.

There are two straightforward rules:

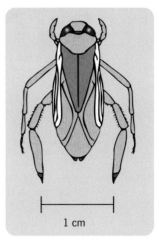

1 cm

Fig. 5 *Freshwater invertebrate (water boatman)*

<table>
<tr><td>**KEY FACT**</td><td>**1** To find the magnification: $\dfrac{\text{size of image}}{\text{size of real object}}$</td></tr>
<tr><td></td><td>**2** To calculate actual size of object: $\dfrac{\text{size of image}}{\text{magnification}}$</td></tr>
</table>

Making dilutions

You may perform some experiments where you must first make dilutions of a given solution. Say you were given a molar solution of glucose (1 M). That would be 1 mol of glucose (180 g) dissolved in 1000 cm³ water. How do you make a 50% solution? Well, obviously it's down to ratios again, this time the amount of water to substance. We would need to make a 1 : 1 mixture of water and the glucose solution. In the same way we could easily make 30% or 70% solutions by diluting in the correct ratio. Look at this table of dilution for making 20 cm³ of the glucose solution.

Concentration of glucose solution/M	Volume of water added/cm³	Volume of 1 M glucose solution added/cm³	Ratio of glucose solution to water
0.30	14	6	1 : 2.3
0.35	13	7	1 : 1.9
0.40	12	8	1 : 1.5
0.45	11	9	1 : 1.2
0.50	10	10	1 : 1
0.60	8	12	1.5 : 1

Can you see how the ratios have been obtained? Obviously, if the volume needed was 30 cm³ the ratio would still be the same. For example, a 0.4 M dilution would still need 1 : 1.5 ratio of glucose solution to water; but, of course, the quantities would be 12 cm³ to 18 cm³.

Chromatography

You may carry out an experiment to separate plant leaf pigments using the technique of **chromatography**, where a solvent carries the pigments different distances along a sheet of filter paper. The distance moved by the solvent from the point of origin to the front is measured and also the distance travelled by each pigment. We can calculate an **R$_f$ value** (the abbreviation comes from **R**atio and solvent **f**ront). So that:

$$\text{The } R_f \text{ value for a pigment} = \frac{\text{distance moved by pigment}}{\text{distance moved by solvent front}}$$

Here are some examples:

Table 2

Name of pigment	R$_f$ value
Xanthophyll	0.51
Carotene	0.96
Chlorophyll a	0.75
Chlorophyll b	0.70

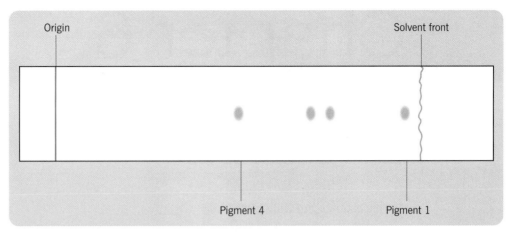

Fig. 6 *Chromatographic separation of leaf pigments*

Exercise 5.3.3

1 Calculate the R_f values of pigments 1 and 4 in *Fig. 6* and identify them from Table 2.

Exam Questions

Exam type question to test understanding of Chapter 5

1 The table shows the ratio of the amount of water lost to the amount of oxygen gained for two terrestrial animals, an annelid worm and an insect.

Organism	Ratio $\dfrac{\text{mass of water lost/mg g}^{-1}\text{min}^{-1}}{\text{volume of oxygen taken up/ cm}^3\text{ g}^{-1}\text{min}^{-1}}$
Annelid worm	2.61
Insect	0.11

Both the annelid and the insect take up oxygen at a rate of $2.5 \text{ cm}^3 \text{ g}^{-1} \text{ min}^{-1}$. Calculate the rate at which water would be lost in the annelid and in the insect.

(Adapted from AQA, Paper 3, summer 1999)

Answers to Test Yourself Questions

Exercise 5.2.1, *p. 54*
1 (a) 2.98 : 1 or 3 : 1
   (b) 3.01 : 1 or 3 : 1
2 3.14 : 1 or 3 : 1 again!
3 red = 697; white = 232

Exercise 5.2.2, *p. 56*
1 2 : 1 : 1 : 2

Exercise 5.3.1, *p. 57*
1 anaphase : metaphase : telophase : prophase : interphase = 1 : 2 : 3 : 14 : 28

Exercise 5.3.2, *p. 58*
1 (a) 2.1%, 4.2%, 6.3%, 29.2% and 58.3% (order as (a))
   (b) interphase   58.3% of 24 h = 14.0 hs
       prophase   29.2% of 24 h = 7.0 h
       telophase   6.3% of 24 h = 1.5 h
       metaphase   4.2% of 24 h = 1.0 h
       anaphase   2.1% of 24 h = 0.5 h
2 (a) clay : sand : humus = 17 : 2 : 1
   (b) 1.8 kg

Exercise 5.3.3, *p. 61*
1 Pigment 1 is xanthophyll and pigment 4 is carotene.

Chapter 6

Displaying data: graphs, charts and scales

In this chapter you will:

- *learn about the types of graph used in biology*
- *find out and learn the rules for drawing graphs*
- *see how graphs can be used to display data from laboratory and field work.*

6.1 Dealing with data

Jamie and Hannah carried out an experiment and drew this diagram of the results:

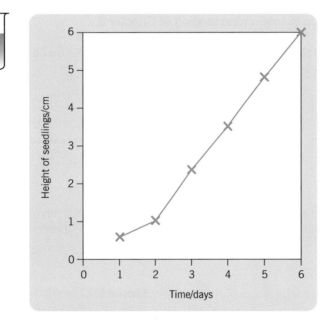

Fig. 1 *Graph to show the relationship between height of wheat seedlings and time*

Well, what could you say about this graph?

1 It is a record of seedling growth.

2 It shows some sort of relationship between height and time.

3 It shows accurate data for only six days.

Wherever you look, in newspapers and journals or at the television, you will come across **graphs** to illustrate information that would otherwise be unintelligible or at the very least boring to assimilate. Graphs make information visually interesting and at the same time, convey much more impact that the raw **data** alone. Graphs are not the sole province of the scientist; you will find economists, weather forecasters and even politicians using this form of presentation to get their message across. Scientists are collectors of information or data. These data are collected systematically and should be presented in suitable **tables** before it is possible to transfer them to charts or graphs. While there is no particular mystique about this scientific method, the presentation of data does have very important rules; in this chapter we will examine the variety of ways data can be displayed. There will be advice on how to avoid some of the common mistakes made by students and even more experienced practitioners.

Choosing the best way to display data

Already in your work in science you will have been carrying out experiments and investigations, and during these you will have collected data. The scientific methods you have employed require you to record these data in a suitable way, for example in tables or **spreadsheets**.

There is a temptation to include too much information in graphs or to present lots of different types of chart, in the hope that one of them might be suitable! This error not only demonstrates a lack of correct analysis, but also suggests a lack of real understanding in the reasons behind drawing graphs.

A word of advice to begin with!

Avoid using every known form of data display in your report! Although it may look attractive, unless the graph is easier to understand than the original table of data, then it is probably not worth drawing.

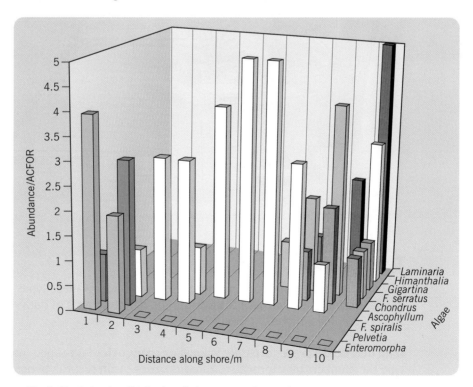

Fig. 2 *Chart showing distribution of algae on a rocky sea shore*

Also, be careful with the use of standard computer graphing packages; they can be very useful indeed, but over complicated if not used wisely. Here is an interesting example. This chart (*Fig. 2*) was copied from a report by a student who thought it looked great.

The chart is hardly easy to interpret is it? It does not represent the nature of the data, which is from a **transect** along a beach. So often students put graphs like this into their work just because they know how to use a program. The trouble is that we are used to similar complex examples in the media, but really data should be presented in the simplest and clearest way.

Types of graph

There are just a few appropriate types of graph or chart that biologists need to use in their work. They include:

● Line graphs

● Bar charts

● Histograms

● Pie charts

● Kite diagrams

● Nomograms

Fig. 3

Types of data

However, before we start to consider the construction of these types of graph, a word about the data itself. It is important to consider the different *types* of data when deciding the best form of graphical representation.

Data may be **discrete** (separate values) or **categorical** (grouped into classes). In the first type the numbers may form a continuous series (as in records of heights of trees) and in the second the numbers may not form a sequence but just be related to numbers of individuals in a category (as in records of eye colour). *Fig. 1*, p. 62 was drawn using two sets of continuous **variables** – time and height. This is fine, but remember that only data from actual records can be manipulated. Averages of estimates should really be avoided; but ecologists often use abundance estimates based on a standard scale, such as the **Domin scale**, for estimating species density, or the 5-point ACFOR scale in *Fig. 2*.

Table 1 *The **Domin scale**, for estimating species density in a set area (1 m^2 quadrat)*

Domin value	% cover
10	91–100
9	76–90
8	51–75
7	34–50
6	26–33
5	11–25
4	4–10
3	Less than 10 (< 10) frequent
2	< 10 sparse
1	< 10 rare

Deciding what sort of graph to use?

It may be obvious from your data that there is a relationship between the two variables chosen. If it is *very* obvious, then it could be argued that you need do no more than present the data in a proper table. If you use a table, consider the best structure. This example shows some of the features of a good table.

Table 2 *Mean composition of human plasma and filtrate*

Molecule or ion	Approximate concentration/g dm^{-3}	
	Plasma	Filtrate
Water	900.0	900.0
Protein	80.0	0.0
Glucose	1.0	1.0
Amino acids	0.5	0.5
Urea	0.3	0.3
Inorganic ions	7.2	7.2

However, if the data in the table seem rather unclear at first or too complex to take in visually, it may be helpful to sketch a quick **scattergram** or **scatterplot** of the raw data, to see what it looks like. An example is shown on the next page.

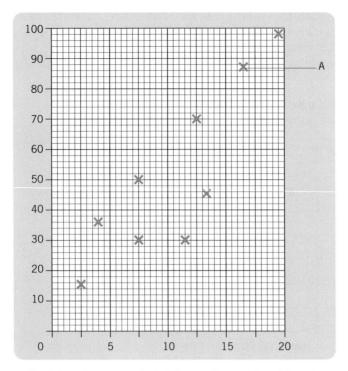

Note

In order to show the methods clearly in print we have used, in graphs, quite a thick 'X' or '●' to show the points to be plotted.

It is strongly advised that in your graphs you use a sharp pencil and to show the *exact* position of the point of crossing use 'X' or '+' or a very small dot with a circle around it – '☉'

Fig. 4 *A scattergram – note that the axes have not been labelled*

Note that each of the points plotted is found using two variables – the co-ordinates – by reading from the scales on the axes. One of the points is labelled A. It has the co-ordinates (17, 87) where 17 is the value on the *x*-axis and 87 on the *y*-axis. You might decide to draw a 'best fit' or **regression line** to pass through the middle of the scatter of the points; we will come back to this later (see p. 106).

Rules of engagement!

Graph questions in exams usually carry a large number of marks for some basic points, so it is really important to get them right. Most of this should be well practised in work done during earlier years, but there is a surprisingly large number of people who still find graphs a bit of a mystery.

Let's start with the piece of graph paper:

● *Which way round should it be?* Choose the best way to accommodate the scales.

● *Where should the axes be placed?* Where there is enough room left to label them.

● *How much of the paper should I use?* Use as much as possible.

● *Do I draw one graph with three curves, or a separate graph for each?*

● *How do I show the data on the graph? Do I use points, crosses, colour, etc.?*

● *How do I join up the data points? Do I use a straight line, a line of best fit or a curve?*

● *Is a title important?* Yes, very!

Fortunately, as you will discover, there is a clear pattern to answer most of these questions. So you should follow the examples shown, but be prepared to amend some of these for special graphs.

6.2 Drawing graphs

Dealing with gaps in data and drawing curves

Look at the results below from an investigation. What seems odd about the information in the table?

A leafy shoot was placed in a potometer, and the volume of water lost by transpiration was measured and recorded. The conditions that the plant was subjected to were changed to see the effects of still or moving air on the rate of transpiration.

Table 3 *Transpiration loss in a leafy shoot*

Time from start of experiment/s	Still air, volume of water lost/mm^3	Moving air, volume of water lost/mm^3
15	24.4	56.3
30	32.5	78.2
60	45.6	121.0
90	56.3	189.3
120	67.4	234.1
180	83.5	357.2

You will notice that the time intervals do not go up in even amounts. It is advisable to collect data over standard intervals of time – say every 30 s. Sometimes this is not possible for some reason (e.g. transpiration may slow down) and so longer intervals are planned to enable sufficient change to take place.

When plotting this information, you must allow for the 'data gap' by drawing an accurate time axis first: using equal increments for every 15 s rather than just putting each item of the **independent variable** (time) in five equally spaced lengths of the *x*-axis (see *Fig. 5*).

You will notice that the curve has been drawn back to the origin, or '0', even though the data in the table does not have any information earlier than 5 s from the start. The only starting point that can be justified from the above table is the 15 s one.

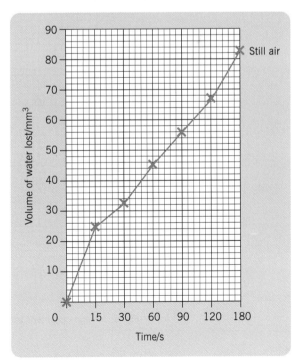

Fig. 5 *What's wrong with this graph?*

Points	We plot these using the co-ordinates e.g. (5, 27).
Graph	This is 'the whole picture', including axes, labelling and title.
Curve	This is the joined points – it's a curve even if it is a straight line or a 'wiggly' line as in Fig. 5.
Axes	These are the scales, which must show what is measured and the units in which they are measured – separated by a '/'.
Origin	This is where the axes meet – it may be at the zero of one, both or neither of the axes.

Look at the data again in Table 3. What is the information telling us? Is it the amount of water lost during each interval or is it the accumulated amount, i.e. has the experimenter measured the amount of water lost for the first interval and added it to the amount for the next interval, and so on?

The answer is that the experimenter has recorded this correctly, by using the cumulative total and not the individual amount each time. If he had then plotted the data, just as he collected it, the resulting graph, *Fig. 6*, would not have made much sense, as you can see. Of course it would have been perfectly reasonable to draw up a table with more columns than the example here, and to put in the loss of water for each interval. Then in another column the cumulative total could be shown. (If this was done in a spreadsheet, one of the columns could have had a formula to do the adding for you.)

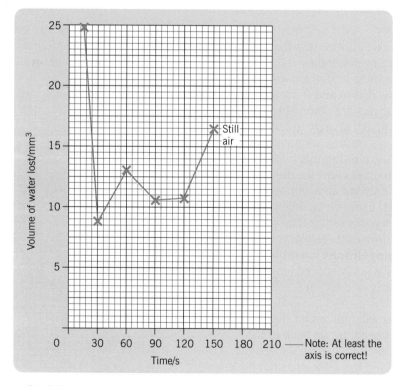

Fig. 6 The experimenter's attempt to graph transpiration loss in a leafy shoot.

This is a good example of how important it is to plan the table that is going to contain the data *before* doing the experiment, so that the data collected are handled correctly.

Remember to think carefully at your planning stage.

Test Yourself

Exercise 6.2.1

1 Some students carried out an investigation to compare the shape of limpets on a sheltered shore with that on a shore exposed to the action of waves. They measured the height (H) and length (L) of 15 limpets on each shore. They then used the ratio of height to length to describe the overall shape of the limpets.

An extract from their field records is shown below.

20/6/94   *Patella vulgata*

SHELTERED SHORE   15 LIMPETS (measurement in cm)

HEIGHT	2.1	2.4	2.6	2.5	2.4	2.7	2.1	
LENGTH	3.6	4.2	4.7	3.9	3.7	4.5	3.7	
RATIO H/L	0.58	0.57	0.55	0.64	0.65	0.60	0.57	

HEIGHT	3.0	2.9	2.9	2.5	2.8	2.9	3.1	2.6
LENGTH	5.2	4.9	5.1	4.6	4.8	5.0	5.3	4.6
RATIO H/L	0.58	0.59	0.57	0.54	0.58	0.58	0.58	0.56

EXPOSED SHORE   15 LIMPETS (measurement in cm)

HEIGHT	1.7	1.8	1.3	2.2	1.9	2.0	1.6	
LENGTH	3.1	3.4	2.9	4.2	4.8	3.8	2.7	
RATIO H/L	0.55	0.53	0.45	0.52	0.39	0.53	0.59	

HEIGHT	1.9	2.0	1.7	2.2	1.9	1.2	1.7	1.9
LENGTH	3.5	4.1	3.5	4.3	3.7	2.8	3.3	4.0
RATIO H/L	0.54	0.49	0.47	0.51	0.51	0.43	0.51	0.48

Prepare a table and organise the data in a suitable way so that the range of shapes of limpets on the two shores can be compared.

(Adapted from Edexcel, Paper T2, 1996)

Another example (below) shows a well-constructed table of data taken from a report written by an Advanced Biology student.

Table 4 *Results for the volume of oxygen generated from catalase acting on hydrogen peroxide at different temperatures*

	Temperature/°C						
	10	15	20	25	30	35	40
Time/min	Volume/cm^3						
1	0.1	0.5	0.5	0.6	1.5	2.5	7
2	0.2	1	1	1.5	2	4	9
3	0.3	2	2	3.2	3	7	11
4	0.4	2.5	2.5	3.9	5	9.5	12
5	0.6	3	3	4.6	6.5	11	13

In the next chapter dealing with manipulating data, you will see more examples of how to interpret data from tables and graphs. Try to look for patterns in the data, and to predict what is likely to happen so that you can spot any anomalous results. Predicting allows you to plan the investigation carefully and to enable you to gather the most useful information.

Graph, bar chart or histogram?

The term 'graph' really covers any form of graphical representation of your data. Two students decided to present the same information in different ways. Look at their graphs.

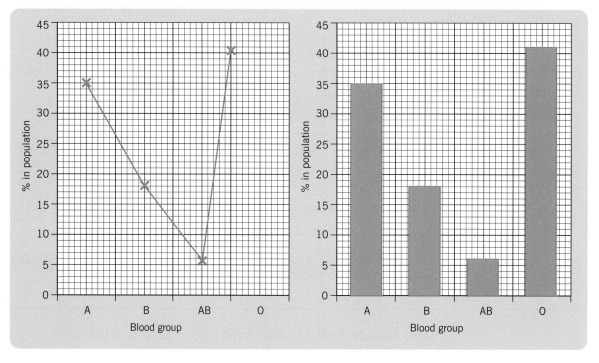

Fig. 7 *Graph showing blood groups in a population*

Clearly, the line graph is inappropriate to display this sort of categoric data. There are only four categories of blood group and being a **discontinuous** variable determined, as you may know, by particular genes, there are no categories other than these four. When the data is discontinuous, the graph is known as a **bar chart** and the blocks are not drawn to touch each other. A line graph could be drawn to show the number of shoppers visiting a supermarket each day for a week. However, if 1000 shoppers were studied every day and the average number of, say, eggs that they bought were our data, then it would be possible to plot that as a bar chart (using blocks of, say, 6, 12, 18 and 24 eggs).

Fig. 8 shows how, when the categories form a continuous sequence, we can draw the blocks touching and that is a **histogram**.

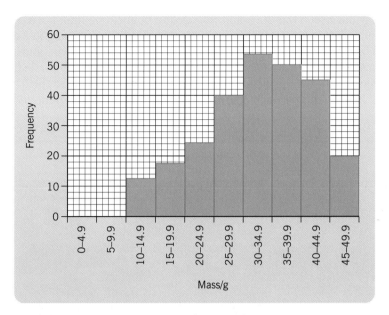

Fig. 8 *Frequency of different mass categories of birds*

Here is another table of results from field work. A particular plant species was examined, and the number of flowers found in 17 quadrats (A–Q) along a transect was recorded.

Quadrat	A	B	C	D	E	F	G	H	I	J	K	L	M	N	O	P	Q
Number of flowers	4	12	21	24	6	27	20	32	8	26	11	18	17	21	15	20	16

The first thing to do in dealing with this data is to devise groups to go along the *x*-axis. Again, the graph drawn is a histogram, since the *x*-axis consists of a continuous variable and the data have been arranged into classes or categories.

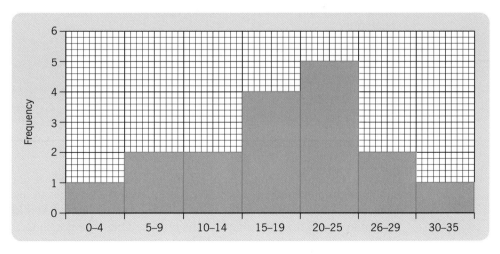

Fig. 9 *Histogram showing numbers of flowers*

You may have seen this sort of graph used in political opinion polls and to display information about population figures or diseases. There are no gaps. Putting in a gap between each bar would be misleading. The limits of each class are clearly shown, and each class is mutually exclusive. The whole area shaded represents the total sample studied, and as the columns are the same width, the heights of the columns are proportional to the frequencies. However, in reality, it is the *area* of each column that truly represents the frequency.

Test Yourself

Exercise 6.2.2

Fig. 10 Holly leaves

1 Harvinder and Tom carried out a short investigation. 24 Holly leaves were picked and their prickles were counted. The number of prickles for each leaf is show below.

16	17	16	16	10	13	17	13
18	17	18	12	16	16	14	5
21	6	13	4	10	22	8	7

Construct a histogram from their data to show the frequency of different numbers of prickles on holly leaves. (Use a tally system for recording the number of leaves in each category.)

HINT > *Group the data into 5 equal categories.*
Plot the categories against number of leaves.

NB Should the bars be separated or next to each other?

There are variations of this type of graph, and one in particular is used extensively in ecological studies. If an area is studied to find out the changes in distribution of certain species along a straight line, the data collected can be plotted as a **kite diagram** or chart. The name comes from the shape of the blocks, which are not square-edged but tail off to the next point. This enables the viewer of the graph to appreciate the *gradual* change or transition between sample points that may be a metre or so apart. Obviously this type of transition is visible on the ground, and so it would be inappropriate to construct sharp-edged block graphs which would suggest a sudden end to the distribution.

How to construct a kite diagram

The data collected here are based on another abundance scale, the **ACFOR scale**, where abundance of plant cover or animal populations is subjectively estimated within a given quadrat. The qualitative letter scale is then converted to a numeric scale, with A (abundant) being the highest at 5, and rare is 1.

Table 5 *The ACFOR scale*

A	Abundant	5
C	Common	4
F	Frequent	3
O	Occasional	2
R	Rare	1
N	Not found	0

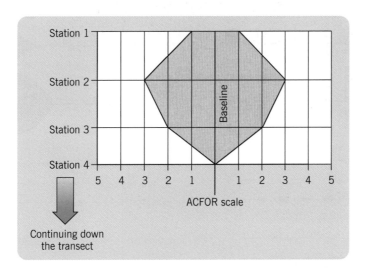

Fig. 11 *Constructing a kite diagram, using the ACFOR scale*

Using the *y*-axis for the species and the *x*-axis for the distance or stations along the transect, the extent of the 'bar' representing each value can be shown as if we were looking down on the transect from above and visualising the extent of the species coverage or abundance.

The kite drawn in *Fig. 11* shows the distribution of a plant which was R in station 1, hence the line starts at 1; F (3) in station 2 and O (2) in station 3. By station 4 the plant was no longer present. Note how the ends of the kite are 'tailed off' to indicate this gradual transition.

A study was made on the distribution of woodland plants. The data here are shown in two forms. *Fig. 12* uses the typical block graph format, whereas *Fig. 13* uses a kite diagram. The latter is regarded as a better way of showing ecological information. As you move along a transect, for example across a path, the vegetation will show change. The tapering ends of the kite-shaped blocks show clearly the gradual variation in this vegetation. The width of the kite represents, in terms of number of graph squares wide, the abundance (from the original ACFOR score) and the length of the bar represents the distance along the chosen transect line.

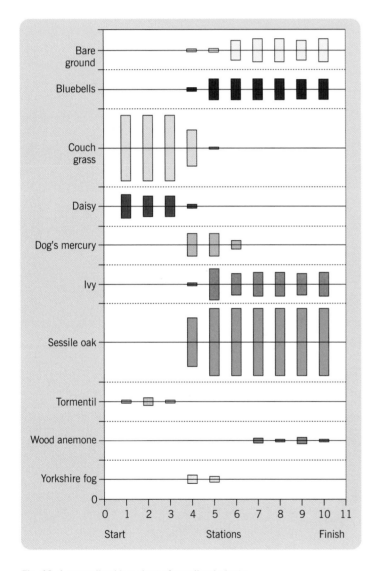

Fig. 12 *A centralised bar chart of woodland plants*

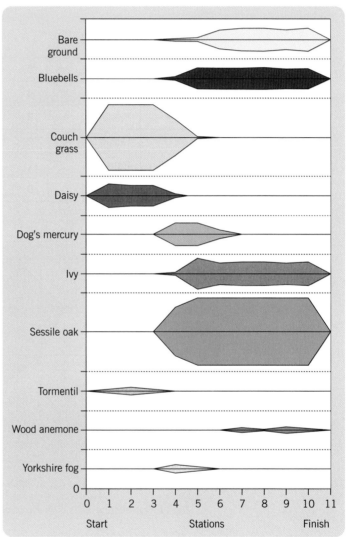

Fig. 13 *A kite diagram of woodland plants*

1 Try producing a kite diagram from the following data, collected by a student and recorded in a field notebook. It's a record of a transect across an area of shingle beach in Devon.

Species	Quadrat								
	1	2	3	4	5	6	7	8	9
Bare ground	A	A	F	C	C	C	R	R	C
Couch grass	R	O	R	R					O
Scentless mayweed			O	R					R
Yellow composite			O	F	R				O
Restharrow			R	O	R		O		
Ribwort plantain					O	R		O	F
Sea carrot					R				
Red fescue grass					C	F	F	A	F
Sea thrift						C	F	R	R

HINT **You must change the letters to their numerical equivalent first.**

6.3 Plotting and drawing line graphs

Let's now consider line graphs.

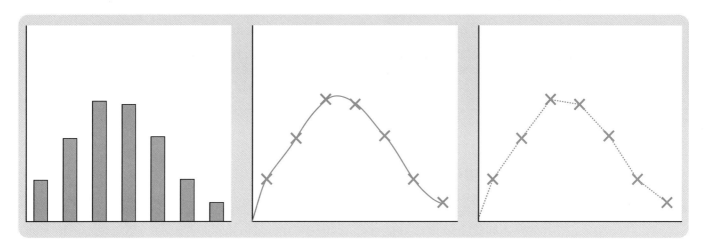

Fig. 14 *Different graphs*

As you can see, there are some subtle differences between the three graphs shown in *Fig. 14*. In each, the values should be plotted and the axes clearly labelled with the quantities from the table of data. The **independent variable** should be plotted on the *x*-axis and the **dependent variable** on the *y*-axis. The independent variables are the intervals that are chosen by the experimenter (e.g. every 30 s). The readings that are observed in the experiment are the dependent variable.

For example, you may want to study the rate of respiration in some germinating mung beans, over a period of an hour. The independent variable here will be the time intervals, which you can choose. As a result of this, you will measure the carbon dioxide output, which will be the dependent variable. You have no control over the dependent variable, as it is dependent on the nature of the organism and its biochemistry and is related to the time interval that you have chosen.

Example

Amber and Paul were on a Pembrokeshire island observing puffins. The parent birds were constantly flying out to sea to catch sand eels and bringing them back to the nest to feed the young birds.

They recorded their observations.
They had decided the time interval, so that will be the *independent* variable and should be plotted on the *x*-axis. The number of eels retrieved is the *dependent* variable (i.e. number depends on time) so that is on the *y*-axis.

Time/h	Number of eels
1	18
2	13
3	15
4	10

If you need help in remembering your axes – use this dreadful mnemonic!

KEY FACT   *Just say 'X is a cross' and 'Y's up!'*

Now let's have a look at putting all this into practice.

Test Yourself
Exercise 6.3.1

1 For plotting graphs of each of these, state which is the independent variable:

(a) date/height (when a child is measured on each birthday)
(b) time/distance (when measuring speed of horse between different landmarks)
(c) volume of oxygen released during ten minutes of a reaction.

Lines on graphs

Look at *Fig. 15*. You will notice that there are four individual sets of information to plot. Often students will draw four separate graphs, each on a new piece of graph paper. This makes it much harder to compare events and trends in the data. It is much better to draw all the separate lines on the *same* axes, but each must be clearly labelled or identified in some way using a key. In textbooks this is sometimes done using different types of line: dotted, dashed, etc. It is difficult to replicate this in hand-drawn graphs; so the best thing to do is to label each line at the end, e.g. with A, B, C and D and then identify these in a suitable box for your key. You can of course use different symbols for the points plotted on each curve or even coloured lines if you wish; although in exams this may slow you down a bit and you could end up with a thick imprecise line.

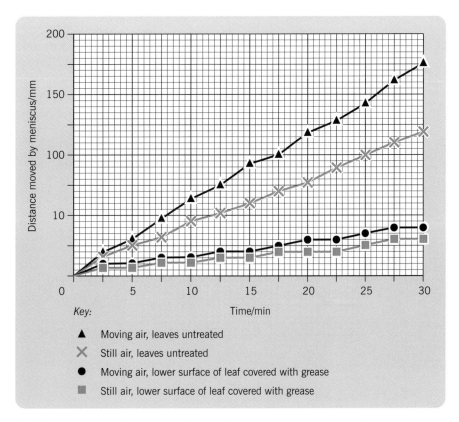

Fig. 15 *Four curves plotted on graph. How would you improve the presentation?*

Points must be clearly marked and accurate, as the co-ordinates are checked by the exam marker. The best way to show them is with a vertical cross, +, as this indicates the intersection of the x/y-co-ordinates most clearly. Do not use a blob (●), which can be inaccurate, or dots which may be indistinct. (Don't forget to use a *sharp* pencil as the magnitude of any error will be directly proportional to the thickness of the pencil line!)

The vertical line of the point can also be extended if there is any information about the error bars or confidence limits (see p. 96). Well, so far, so good!

Now how *are* the points to be joined up? This may seem to be very simple, but it is surprising how controversial this has become!

RULE 1   *If the data show a continuous relationship between the two variables then the curve can be a straight line of best fit, joining or approximating to the points. Usually in these cases the **gradient** or other relationship can be calculated (p. 95).*

RULE 2   *In most biological examples there is no such continuous relationship. In these cases the biological convention is that straight lines should connect each point to the next one. A smooth rounded curve can only be justified if there is good reason to think that the intermediate values fall on the interpolated line. Straight lines between points show that the accurately plotted recorded points are fixed, but in between them the values are unknown and cannot be predicted.*

RULE 3   *In other disciplines, the line of best fit is often taken as the norm, since data are predictable. For example, in physics, you could plot a velocity/time graph like this:*

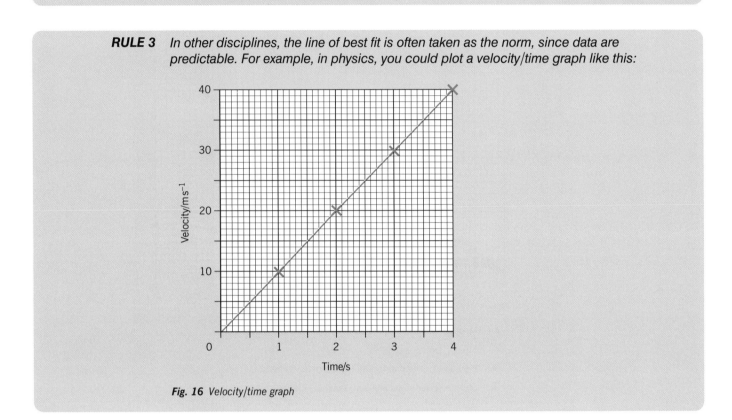

Fig. 16 *Velocity/time graph*

Example

In a graph of a biological reaction, you may have only five points on your graph as in *Fig. 17* below. How many different curves could you draw which would link these points? Could you be certain of any one of them?

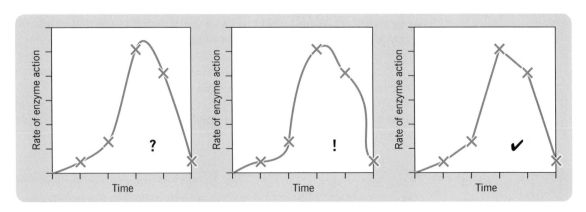

Fig. 17 *Any one of these could be drawn, but the only sure way is to join up all known data plots with straight lines*

6.4 Other types of graph

Pie charts

Pie charts are very useful when the data are discrete (categories). A circle is divided according to the proportion of counts in each category. They help when there are up to about six categories, but can become confusing if you use them for more.

Example

The following skulls were found in barn owl pellets:

Skulls	Number in pellets
Field mouse	23
Bank vole	8
Yellow-necked field mouse	2
Shrew	7
House mouse	3

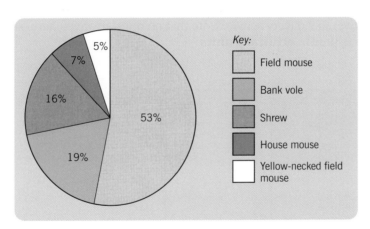

Fig. 18 *A pie chart showing the proportion of mammal skulls found in samples of barn owl pellets*

Test Yourself

Exercise 6.4.1

1 Draw a pie chart for these data:

Blood group	Percentage of UK population	Number of degrees of circle
A	46	$\frac{46}{100} \times 360° = 166°$
B	8	$\frac{8}{100} \times 360° = 29°$
AB	2	$\frac{2}{100} \times 360° = 7°$
O	44	$\frac{44}{100} \times 360° = 158°$

HINT

You surely remember that the whole circle has 360°, the half circle 180° and the quarter circle segment has 90°. So when you know the percentage of the population in a certain category, you can work out how many degrees of the 'pie' – the complete circle – that category should occupy.

Now all that you have to do is to use a pair of compasses to draw a circle and a protractor to measure the angles. The convention is to start with the 12 o'clock position and to move in a clockwise direction; draw first the biggest angle, in this case 166° and then in descending size order, the remaining angles. Having drawn the pie chart, label the segments and provided a key.

The nomogram

The **nomogram** is an unusual method of graphic presentation of data. It is used when three variables can be related to each other.

The nomogram in *Fig. 19* can be used to determine the energy needs of males aged 10 to 15. The left-hand line shows a range of body mass from 30–65 kg and the right-hand line gives an estimate of how active the person is. The range of physical activity level (PAL) goes from 1.4 (inactive) to 2.0 (very active). Note that no units are given here, so we can say that the PAL is measured in **arbitrary units**.

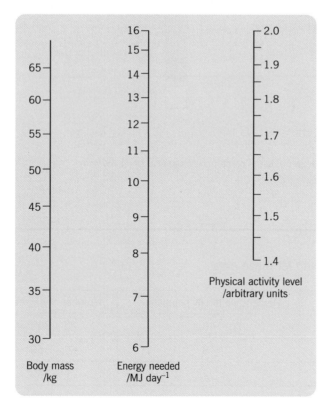

Fig. 19 *A nomogram for physical activity (for males aged 10–15)*

Example

The first calculation to attempt is for an inactive boy (PAL = 1.4) with a body mass of 62.5 kg.

Method

You need to use a straight-edge (rule) to join the two co-ordinates (1.4, 62.5 kg).

Where this line crosses the middle line you can make a direct reading of the energy requirement of this person. It is 10.3 MJ day^{-1}.

Example

Now, what about an active person (PAL = 1.9) and a body mass of 50 kg? Well, use the same process and you should find an energy need of 12.2 MJ day^{-1}.

There is another example of an exam type nomogram on p. 83.

HINT

> *Questions of this sort (as well as a number of other graph questions) often ask you to do a straightforward reading and then to write brief answers to sub-questions around the topic. So an exam question might include the nomogram above and you might then be asked to make comparisons between the energy requirements of a pregnant person, a baby and a sedentary pensioner, for example.*

Summary

Now that you are more aware of the variation that exists in graph drawing, you will be able to experiment with different ways of displaying your data. Just remember the following simple rules:

KEY RULES

1 *Decide, at the start, which type of graph you intend to use.*
2 *Collect data that will be relevant to the type of graph you have chosen.*
3 *Select appropriate scales for your graph that fit the graph paper adequately.*
4 *Always label your axes, make sure there is a key and always put a title.*

If you look at the typical mark scheme that an examiner uses for marking your graphs, you will see:

A = axes round the right way/correctly labelled
S = scale suitable, more than $\frac{1}{2}$ paper used
P = points accurately plotted
C = points joined up appropriately, with either straight lines or curvy line.

Exam Questions

Exam type questions to test understanding of Chapter 6

1 An experiment was carried out to investigate water loss from two arthropods, woodlice and caterpillars (of the large white butterfly).
Ten woodlice and ten caterpillars were exposed to a range of temperatures, from 10 °C to 60 °C, in dry air. The evaporation of water from the surface of each animal was measured in mg cm^{-2} h^{-1}.

The results are shown in the table below.

Temperature/°C	Rate of evaporation of water/mg cm⁻² h⁻¹	
	Woodlice	Caterpillars
10	5.0	0.0
20	7.5	0.0
30	10.0	2.0
40	14.5	3.5
50	19.0	9.0
60	26.0	13.5

Plot these results on graph paper.

HINT

Remember to use the rules that you have practised. There may be 5 marks for this type of question.

(Adapted from Edexcel, Paper B6, 1997)

2 *Selenastrum capricornutum* is a single-celled green alga which grows in freshwater. Cultures of *S. capricornutum* produce green suspensions and the growth of this alga can be measured using a **colorimeter**. A colorimeter is an instrument which measures the amount of light absorbed by a coloured suspension. As the number of cells increases the absorbance increases.

An investigation was carried out to compare the growth of *S. capricornutum* in two samples (A and B) of water from a river. Sample A was taken upstream of a drain into the river from agricultural fields. Sample B was taken downstream from the drain. Both samples were filtered, then each was inoculated with the algae. The cultures wer   pt at a constant temperature and constant light intensity and the absorbance of each cul   e was measured daily for seven days. The results are shown in the table below.

Day	Absorbance/arbitrary units	
	Sample A	Sample B
0 (Start)	0.00	0.00
1	0.01	0.07
2	0.23	0.42
3	0.35	0.64
4	0.38	0.78
5	0.38	0.88
6	0.38	0.94
7	0.38	0.97

Plot these results on a piece of graph paper.

(Adapted from Edexcel, Paper B6, 1998)

3 The nomogram below (*Fig. 20*) was produced by a research group of sports scientists for a national team of women athletes. The left-hand line shows the Groups A–E classified according to the distance of their specialist event. The right-hand line shows the speeds that they were instructed to run at before a blood sample was taken. The lactate level in the blood was measured, because it is known that at a blood lactate level of less than 4.0 mmol dm^{-3} the athlete is respiring aerobically. Above that level the athlete is respiring anaerobically to get the energy required.

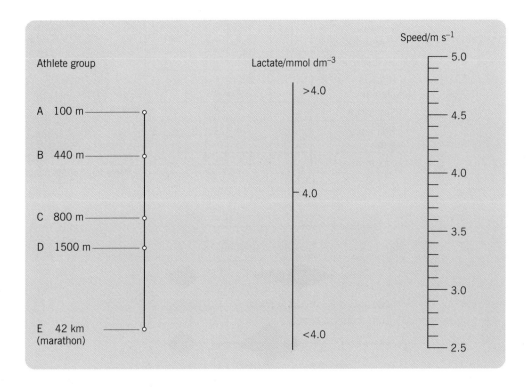

Fig. 20 *Data for athlete groups from 100 m to 42 km (for women)*

(a) At what speed must the group of 800 m athletes run to have a lactate level of 4.0 mmol dm^{-3}?

(b) Which of the athletes could run at 4.4 m s^{-1} with aerobic respiration?

(c) Predict, with a reason, whether or not 200 m athletes running at 3.8 ms^{-1} would be respiring aerobically.

Answers to Test Yourself Questions

Exercise 6.2.1, *p. 69*

1 The data should be organised with: correct table of raw data (height, length, ratio) for the sheltered shore; correct table for raw data (height, length, ratio) for the exposed shore. The mean ratio of sheltered is 0.58 and the mean of exposed is 0.5. The figures should be tallied into suitable size classes for each shore.

Exercise 6.2.2, *p. 72*

1

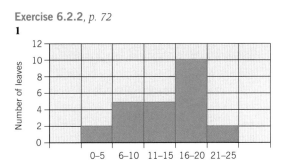

Graph showing number of prickles on holly leaves

Exercise **6.2.3**, *p. 75*

Transect across shingle beach in Devon

	Quadrat								
Species	1	2	3	4	5	6	7	8	9
Bare ground	5	5	3	4	4	4	1	1	4
Couch grass	1	2	1	1					2
Scentless mayweed		2		1					1
Yellow composite			2	3	1				2
Restharrow			1	2	1		2		
Ribwort plantain					2	1		2	3
Sea carrot					1				
Red fescue grass					4	3	3	5	3
Sea thrift						4	3	1	1

Exercise **6.3.1**, *p. 76*

1 The independent variables are:
  (a) date
  (b) distance
  (c) time

Exercise **6.4.1**, *p. 79*

1

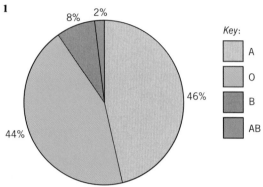

Percentage of UK population blood groups

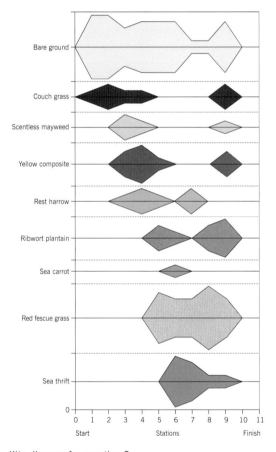

Kite diagram for question C

Chapter 7

Using graphs and interpreting data

After completing this chapter you should be able to:

- *use the skills of drawing and interpreting graphs*
- *find values by interpolation and extrapolation*
- *understand measuring gradient and calculating rates*
- *use logarithmic graphs*
- *introduce error bars.*

7.1 Interpreting graphs

After *producing* the graph from data collected in experiments and investigations, what else can we do? Well, one of the most important things you are expected to do is to interpret given graphs. Consider this example of a graph showing certain data.

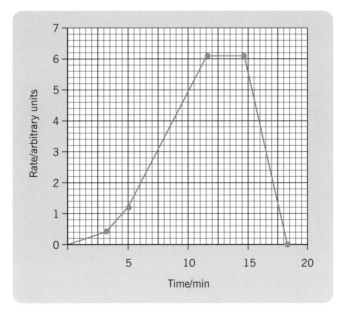

Fig. 1

Question
Comment on the shape of the graph.

Answer

This is what could be a suitable answer:

During the first 3 min the rate increases slowly, then over the next 2 min the rate increases more. After this, there is a steep increase in rate until a maximum rate of 6.0 units is indicated at 11.5 min. The rate remains constant for 3 min and steeply declines to zero at 18 min at a constant rate of decline (a straight line).

HINT

In your explanation of what a graph shows don't just state what is happening to the curve, be sure also to make use of the numerical data from the graph, showing that you understand the graph

7.2 Calculations from graphs

Apart from the act of drawing graphs, there is another important aspect to the process. A series of numbers in a table can show some form of trend or pattern, but this becomes much clearer when plotted in a graphical way. Finding out the rate of change over a period of time is also much simpler from a graph than from the raw data.

Here are some data from an experiment on protein digestion. They come from an exam paper that asks you to do three things: 1. plot, 2. comment and 3. calculate. The best strategy in an exam is to read through the whole question (and in this case note that you do not really need to understand the biology to get any, except one, of the marks!)

Question

An experiment was carried out to determine the effect of temperature on the activity of an enzyme digesting the protein gelatin.

Gelatin was incubated with the enzyme at a range of temperatures from 5 °C to 60 °C. The rate of amino acid production was measured over a three-hour period. The results are shown in the table below, expressed as rate of amino acid production in mg dm^{-3} h^{-1}.

Temperature/°C	Rate of production of amino acid /mg dm^{-3} h^{-1}
5	14
10	19
15	24
20	31
25	40
30	51
35	68
40	93
45	98
50	89
60	33

1 Plot the data on graph paper. *(4 marks)*

2 Comment on the effect of temperature on the activity of the enzyme as shown in the graph.
(**3 marks**)

3 The experiment at 45 °C was continued for a further 7 h. At the end of this time, an additional 292 mg dm^{-3} of amino acid had accumulated. Calculate the mean rate of production during the 10 h at 45 °C. (**3 marks**)

(Question and answer scheme adapted from Edexcel, Paper B6, Synoptic paper, 1996)

Answer

1

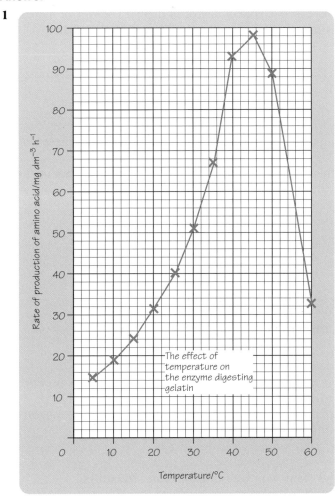

Fig. 2 *Note that the graph is given a title and the axes are labelled*

HINT ⟩ *The points are joined up with straight lines, and each point is shown with a cross at the co-ordinates. Marks are also given for labels on axes, correct plotting of points and a title. (Did you notice that there is no reading for 55 °C?)*

Now for the second part of the question:

2 Comment on the effect of temperature on the activity of the enzyme as shown in the graph.
(**3 marks**)

'Comment' here means simply that you should refer to the data and describe what is happening during the time period of the graph. There is no need to go into any explanation at this stage. What is important is to use data from the graph itself. (Note the use of temperature and rate readings from the graph in the following mark scheme.)

Marks are given for:
   the rate increased exponentially/or until 40 °C; slower/or increase to
   45 °C/maximum/optimum rate at 45 °C/reaches a peak at 45 °C; (note that the / indicates alternatives). *Any of these points will score the 1 mark.*

   reference to steep/e.g. fall/reference to specified figures/after 45 °C/50 °C/after optimum/or correct reference to specified figures and temperatures. *Any of these will score 1 mark.*

   reference to (kinetic) energy/movement of molecules/given example; reference to denaturing of enzymes at higher temperatures. *Any of these will score 1 mark.*

(Note that this last mark is the only one in the question that needs some *biological knowledge* – all of the other marks are for graph skills.)

The last part of the question needs a calculation:

3 The experiment at 45 °C was continued for a further 7 h. At the end of this time, an additional 292 mg dm^{-3} of amino acid had accumulated. Calculate the mean rate of production during the 10 h at 45 °C. **(3 marks)**

The answer is:

   [(3 × 98) + 292] ÷ 10 = 58.6 or 59 mg dm^{-3} h^{-1}

You know that at 45 °C, 98 mg dm^{-3} of amino acid is produced in each hour; So 3 × 98 mg dm^{-3} (=294 mg dm^{-3}) is produced in 3 h. Add to this 294 mg dm^{-3} the next 7 h amount of 292 mg dm^{-3}, which gives 586 mg dm^{-3} produced in 10 h. Divide by 10 to obtain the rate for 1 h.

NB Always show your working and state the units.

Marks are given for:

   the correct method (1)
   the correct answer (1)
   the correct units (1).

It is easy to tap away on your calculator and then just put the final answer down on your exam paper. Remember the examiner wants to see how you came to that answer, so show your data working clearly. You may still get 1 mark for the correct method even if you get the answer wrong.

7.3 Logarithmic scales

Sometimes in experimental work, the data collected are not easy to plot on a graph because of the very wide range of numbers involved. A typical example comes from measuring the growth rate of microbes, such as yeast in a culture. You may carry out a laboratory exercise where the cultures are set up, and samples of yeast cells to be counted are removed; a special slide known as a haemocytometer is then used to estimate the number of cells in a specified volume of culture liquid (see p. 26).

This process is repeated over several days. The results that you are likely to get could be plotted and may look something like *Fig. 3*:

Fig. 3 *Graph of cell number against time*

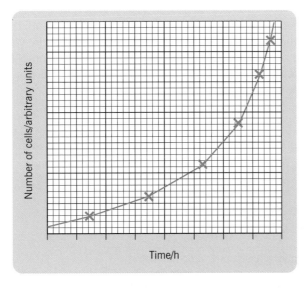

Fig. 4 *Graph of yeast population growth (plotted on logarithmic paper)*

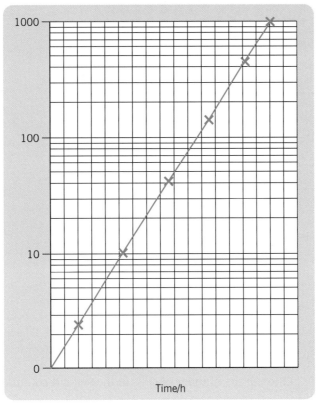

In the graph (Fig. 3) the data form a curved line; if the experiment had carried on for longer, the y-axis would have become extremely long! This shape with increasing steepness is described as **exponential**. It is difficult to read in the steep part so there is a way of getting around the problem. The same data are plotted in *Fig. 4*.

In *Fig. 4*, you will see that the y-axis is very different. The horizontal grid lines occur in three blocks. Each block represents a part of the logarithmic scale ($\log_{10}$). The first block goes from 1–10, the second *equal-sized* block from 10–100 and the third from 100–1000. If we were to extend the y-scale further the next equal-sized block would be 1000–10 000 and so on. In this way the scale becomes manageable for exponentially increasing data. (Note also that the x-axis in this example is the standard linear arithmetic scale.)

Logarithmic-arithmetic paper (as used in *Fig. 4*) is not easy to find; there is another method, which is more commonly used. Rather that plotting the arithmetic value (that is, the actual raw data) it is better to find the log to the base ten of the numbers, i.e. to convert the figures to a **logarithmic scale**. Look at the increase in number of cells in a culture over a period of 15 h, shown in Table 1.

HINT *To find $\log_{10}$ of a number (e.g. 16) key in:* log 16 = . *The answer should be 1.204. In Table 1, these log calculations have been done to two decimal places – check them!*

Table 1

Hours (h)	Number of cells	Log number of cells ($\log_{10}$)
1	1	0
2	2	0.30
3	4	0.60
4	8	0.90
5	16	1.20
6	32	1.51
7	64	1.81
8	128	2.11
9	256	2.41
10	512	2.71
11	1024	3.01
12	2048	3.31
13	4096	3.61
14	8192	3.91
15	16 384	4.21

You will see that, although the number of cells in the second column is increasing exponentially and would be very hard to plot as a standard graph, the $\log_{10}$ of the numbers are much more manageable.

You will have to decide how many decimal places to include for the figures.

Test Yourself

1 The table contains data on the number of cases of AIDS and HIV infections in men by age. Plot these data as $\log_{10}$ against age.

Exercise 7.3.1

Age	0–4	5–9	10–14	15–19	20–4	25–9	30–4	35–9	40–4	45–9	50–4	55–9	60–4	65+
Number	306	148	183	747	3989	7517	7561	5623	3501	2105	1237	691	392	240

2 Look at this graph showing the number of yeast cells produced over several days.

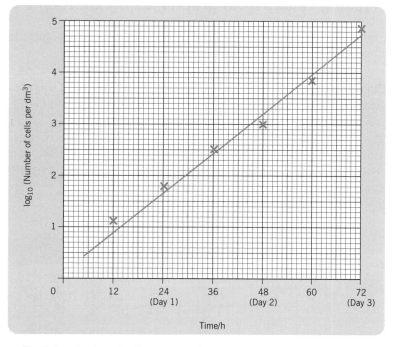

Fig. 5 *Growth of yeast cells over several days*

(a) From the graph, calculate the actual number of cells in the culture after 54 hours.
(b) What is the growth rate per hour during the period of time between 24 and 36 hours?

HINT *Don't forget to include the units!*

7.4. Using graphs

You have come across many types of graph, but in examination questions you will also have to interpret information from graphs. This is, after all, what graphs are about! So try to interpret in your own words some of graphs in the book, and check after you have written your *full answers*.

HINT *Always use numbers and other information that are in the graph to answer. Avoid vague comments like 'after a while the evaporation rate slowed down'; it would be much better to write, 'after the first 10 min of the experiment the rate of transpiration decreased from 14 mm³ min⁻¹ to 6 mm³ min⁻¹.'*

This section includes more hints and suggestions in using graphs and charts. Although you may not have to carry out all of these tasks, it is a good idea to be familiar with them, especially if you are doing investigative work of your own.

Remember that it is not a good idea to include too many ways of interpreting data; choose the most appropriate for the task.

Reading values from graphs, using intercepts and co-ordinates

You should already be familiar with reading values from graphs, but there are some simple hints here to help you to find your way around.

The **intercept** is the place where a graph line cuts one of the axes of the graph. The value of this can be read directly on the scale on the axis (see p. 108).

The **co-ordinate** is the value that represents the position of the point on a graph where x-and y-data meet. They are in fact the points that are plotted in the first place. Imagine it as reading an Ordnance Survey map and trying to use a grid reference to locate a particular spot.

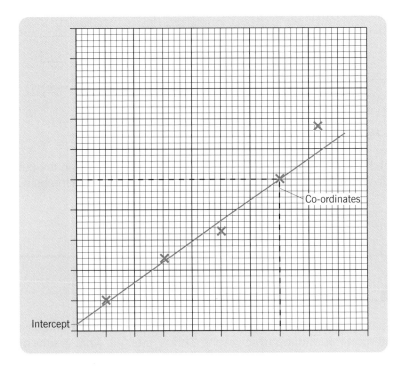

Fig. 6 *Identifying co-ordinates and intercepts*

Interpolation

Interpolation means determining a value, not plotted but within the range of the experimental data, by reading directly from the graph or by calculation.

Example
If you carried out the following experiment on osmosis, using just a few samples, it would be possible to interpolate a value for the osmotic concentration of an unknown sample.

A class was presented with 5 solutions of sucrose of differing concentrations. They found the osmotic pressure produced as shown in the table of results.

They plotted these figures on a simple graph making the *x*-axis show concentration and the *y*-axis, the osmotic pressure. They were asked to find from the graph, the osmotic pressure produced by a 0.25 M sucrose solution.

Concentration of sucrose/M	Osmotic pressure/kPa
0.10	258
0.20	542
0.30	818
0.40	1118
0.50	1452

To find this value, after plotting the graph it is a simple matter of reading up from 0.25 M on the *x*-axis and across to read the value of the intercept on the *y*-axis (as in *Fig. 7*). The value is 680 kPa.

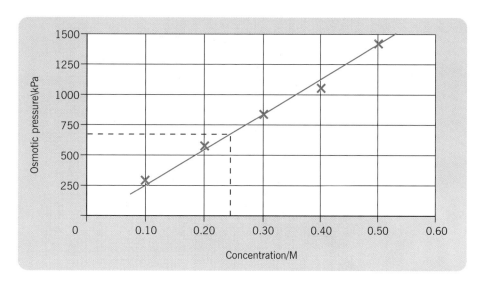

Fig. 7 *Osmotic pressure produced by sucrose solutions*

Extrapolation

Extrapolation means determining a value that is outside the range of experimental data, by extending a graph or by calculation. The assumption is that the trend of the data continues unchanged beyond the experimental region of the graph. This is a useful tool to answer the 'What if' questions in science.

Example

An experiment was carried out to investigate the changes in concentration of carbohydrates and lipids in the blood of a locust during a long period of flight.

Measurements were made of the amount of monosaccharide (carbohydrate) and diglyceride (lipid) at the beginning of the flight and at 60 min intervals during the flight, which lasted 300 min. The results are shown in the table below.

Time during flight /min	Concentration of monosaccharide/μm mm^{-3}	Concentration of diglyceride/μg mm^{-3}
0	30.0	3.0
60	13.0	10.0
120	12.0	19.0
180	11.5	20.0
240	11.0	20.0
300	11.0	20.0

(Data adapted from Edexcel)

We can plot these data on graph paper (*Fig. 8*).

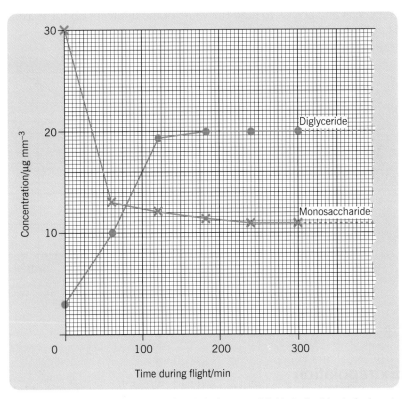

Fig. 8 *Changes in concentration of carbohydrates and lipids in the blood of a locust during a long flight – showing possible extrapolation*

If we were to extrapolate from this graph, it would be fair to deduce that the rate of change in both diglyceride and monosaccharide concentration would allow us to continue the line if we were to extend the experiment for, say, a further 25 min. However, we would not be so confident if the curves were more erratic.

Extrapolation is only possible if there is a clear trend in the data.

HINT *One word of warning: be careful about extrapolating curves in biology, particularly with enzyme-controlled reactions. There is a tendency for the reaction to slow down or decrease rapidly after a general increase. This may be because of the denaturing of the enzyme.*

Measuring the gradient of a graph

Straight line graphs

The **gradient** of a straight line in a graph is the *change* in the *y*-value of a graph divided by the *change* in *x*-value. This is sometimes said as 'rise over run' (see *Fig. 9* and p. 108).

Why measure the gradient? It can enable the scientist to calculate the rate of a reaction in biology – for example, in an enzyme-controlled reaction, the points can be plotted on a graph and a curve drawn. The rate of enzyme activity can be calculated from two points on the line – for a period of time and the other for the amount of product (or disappearance of substrate).

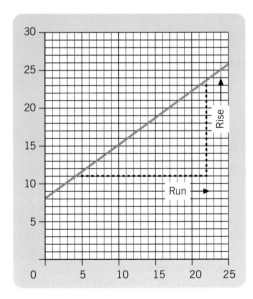

Fig. 9 A gradient

KEY FACT *To calculate the gradient of a straight line:*

- *Determine the values of change in y(Δy = rise) and changes in x(Δx = run).*
- *Calculate the gradient $\Delta y / \Delta x$ (rise/run) (see p. 108).*

Curved graph

If you have to find the gradient for a point on a rounded curve (rather than a straight line), then first identify a part that is significant (as the gradient of a curve is obviously changing).

> *KEY FACT* To calculate the gradient of a curve:
>
> - *Select the point at which you want to measure the gradient on the curve. You might want to select a particular time, if that is relevant.*
> - *Draw a tangent at that point (i.e. a straight line that just 'brushes' the curve at its mid-point).*
>
> *Fig. 10* A tangent of a curve
>
>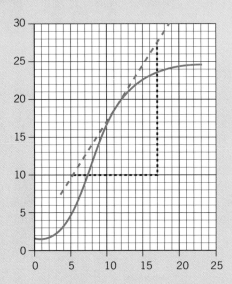
>
> - *Determine the values of Δy and Δx in the same way as for the straight line. Calculate the gradient $\dfrac{\Delta y}{\Delta x}$ in the same way.*

Showing levels of error on graphs

This doesn't suggest that you have made a mistake – error lines are lines drawn on a graph to show the *uncertainties* in the experimental data.

In carrying out experimental work, it is not possible to be 100% accurate in what you do. Measurements using instruments such as thermometers, pH meters and colorimeters rely on the *precision* of the observer and the *calibration* of the instrument. You may, for example, be measuring coleoptile lengths of wheat seedlings and record one as 4.5 cm. The trouble is that you can only be sure of the length to within 1 mm. You know that it is not as low as 4.4 nor as high as 4.6 – but there is an uncertainty. You would record this as 4.5 cm ± 0.05 cm or 45 mm ± 0.5 mm (this is read: '45 mm plus or minus 0.5 mm'). This uncertainty (or **error** – though it's not your fault!) can be indicated by drawing an **error bar** above and below the plotted point on the graph, as shown in *Fig. 11*.

Similarly, as all good scientists know, it is advisable to repeat experiments to obtain additional data, which will show you the mean result. Calculation of the **standard** deviation of the mean for each data point (see p. 123) can then be shown on a graph. This is useful to indicate that you are aware of the limitations of the technique or sampling method, and it suggests that you have taken the data a little more seriously!

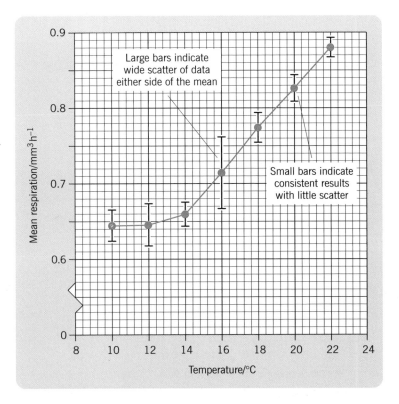

Fig. 11 *Effect of temperature on respiratory rate in germinating Mung beans*

Example
Jenny was studying the effects of varying nitrate concentrations on the growth of algal cells, as part of a eutrophication study. She obtained the following data after 28 days.

Table 4 *Algal cell counts at differing nitrate levels*

Calcium nitrate concentration/g dm^{-3}	10.0	1.0	0.4	0.1	0.01	0.001	0.0001	0.0
Mean cell number $\times 10^{-5}$	5.43	5.00	6.63	9.75	11.53	6.73	5.30	5.50
Standard deviation $\times 10^{-5}$	±0.55	±0.44	1.17±	±1.20	±2.12	±1.13	±0.46	±0.42

Fig. 12 shows the graph with the standard deviations given as error bars.

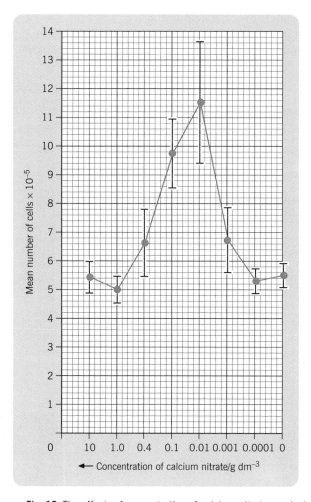

Fig. 12 *The effects of concentration of calcium nitrate on algal cell growth*

Exam Questions

Exam type questions to test understanding of Chapter 7

1 An investigation was carried out into the effect of light intensity on the uptake and release of carbon dioxide by two green plant species J and K.

Single leaves of each species, still attached to plants, were sealed in glass vessels containing a known mass of carbon dioxide. The leaves in the vessels were exposed to light of known intensity for one hour. The change in mass of carbon dioxide in each vessel was determined and the change in mass of carbon dioxide per square centimetre of leaf surface was then calculated. The experiment was repeated at a range of light intensities for both species.

The results are shown in *Fig. 13*. Light intensity is expressed as a percentage of normal daylight.

(a) (i) From the graph, determine the light intensity at which there is no net exchange of carbon dioxide by leaves of species J and K. **(2 marks)**

   (ii) Comment on the relationship between light intensity and the exchange of carbon dioxide in species J. **(3 marks)**

(b) (i) Give *two* differences between the curves for species J and K. **(2 marks)**

(Adapted from AQA, BY06, March 1999)

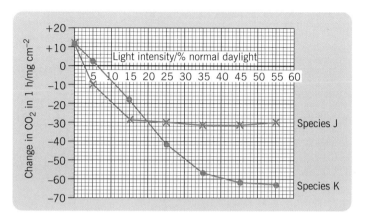

Fig. 13 *Change in CO_2 release in two green plants*
(Adapted from Salisbury and Ross (1978), Plant Physiology)

2 The yeast, *Candida utilis*, was grown in a liquid culture medium. Every hour, a 1 cm^3 sample was taken from the culture and diluted 100 times. One drop of the resulting suspension was then placed on a haemocytometer slide. *Fig. 14* shows the mean number of yeast cells in each 0.004 mm^3 of the haemocytometer grid at different times after the yeast was added to the medium.

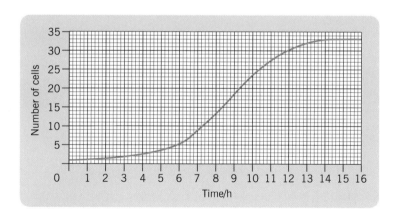

Fig. 14 *Growth of yeast cells*

Use the graph to calculate the number of yeast cells per cubic centimetre in the yeast culture at 13 h. Show your working.

(Adapted from AQA, BY06, March 1999)

Exercise 7.3.1, *p. 91*

1

Age	0–4	5–9	10–14	15–19	20–4	25–9	30–4
Log_{10}	2.49	2.17	2.26	2.87	3.60	3.88	3.88
Age	35–9	40–4	45–9	50–4	55–9	60–4	65+
Log_{10}	3.75	3.54	3.32	3.09	2.84	2.59	2.38

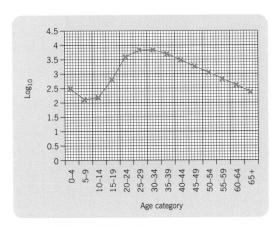

Cases of AIDS and HIV infections

2 (a) 3.6 cells on log_{10} scale to convert this to the actual number of cells key

and the answer is 3981 cells dm^{-3}

(b) Growth from 1.7 to 2.5 on log_{10} scale or in actual numbers from 50–316 cells in 1 dm^3 of culture solution in 12 hours, i.e. 22 cells dm^{-3} h^{-1}

Chapter 8

Correlation and regression

After completing this chapter you should be able to:

- *understand positive, negative and nil correlation*
- *use scattergrams for ecological and laboratory data*
- *calculate regression*
- *use the statistical r_s as a correlation index.*

8.1 Correlation

Imagine test data collected from a group of eight 9-year olds:

Name	English mark	Maths mark
Andy	90	87
Brian	85	83
Cathy	80	84
David	72	68
Elaine	70	62
Fiona	67	66
Gill	61	64
Harry	58	61

You could plot a graph of these marks. There would be no real reason for putting either variable on the *x*-axis, so your scatter of points may look like *Fig. 1*.

Fig. 1 Scatter of marks (positive relationship) (Note that this graph does not use the 0,0 point of origin)

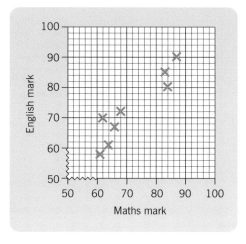

There is obviously a relationship between the abilities of these children in the two subjects. Those who do better in one subject, also achieve more in the other. We say that there is a **positive relationship**. As one increases, so does the other.

Possibly in another pair of subjects the situation could be very different. The better pupils in one subject could be the least able in another. That situation may appear in a graph like *Fig. 2*

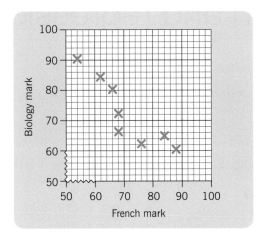

Fig. 2 *Scatter of second set of marks (negative relationship)*

That demonstrates a **negative relationship**. You might expect to see a similar relationship if you plotted information from a number of countries where an increase in the standard of living showed a decrease in the rate of infectious disease. Is the graph (*Fig. 3*) an example of a negative or positive relationship?

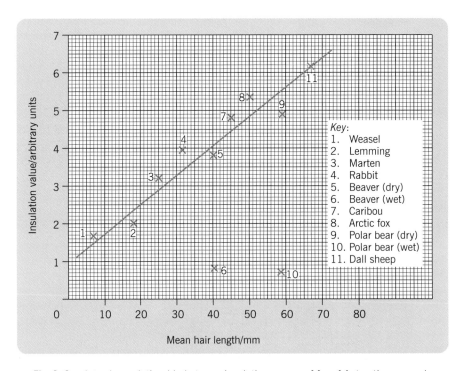

Fig. 3 *Graph to show relationship between insulating powers of fur of Antarctic mammals and length of hair (Pough et al., 1996)*

Using 'invented data', such as in *Figs 1* and *2*, it is easy to make a relationship look as though all of the points are on, or close to, a straight line. This type of graph is a **scattergram** (or scatter diagram or scatter plot). The labelled scales do not have to start at zero, one does not depend on the other; it must also be realised that they only demonstrate a **tendency**. In the next two graphs (*Fig. 3* and *Fig. 4*) the relationships are not perfect but are said to show a **positive correlation**. In *Fig. 3* there is a strong positive correlation. Although in *Fig. 4* the points are scattered rather further from the line, there is still a positive correlation, fairly close and within a band (from 'bottom left' to 'top right') so showing a clear tendency.

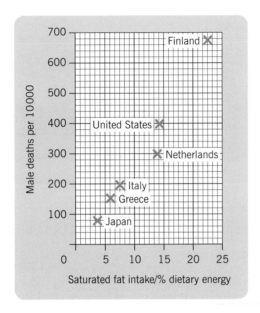

Fig. 4 *Graph to show relationship between intake of fat and male death rates (Taylor, 1995)*

Sometimes there is **nil correlation** and the plotted points are scattered all over the area between the axes!

Note here a common pitfall – some wrongly assume that they can state from a positive correlation that one factor *causes* the other. It may be tempting to argue from *Fig. 4* that a high intake of saturated fats causes a high death-rate. That can't be assumed – there could be other explanations (genetic, environmental or physiological). You must not assume a **causal** relationship – that either one is caused by the other. In the early days of epidemiology, scattergrams seemed to indicate a close positive relationship between lung cancer and the number of cigarettes smoked. However, the medical researchers started by discounting the apparent link; they thought it could have been a spurious association and only used it as an indication that further investigations should be set up.

Looking for relationships in data from investigations

After carrying out an investigation where we have recorded data that may be related, we need to test to see if there is a convincing relationship between the **independent variable** (the input – that which the experimenter selects and sometimes called the explanatory variable) and the **dependent variable** (the output – the variable that depends on the input and sometimes called the response variable). The results from a student's laboratory work follow.

Mwamba carried out an investigation into the effects of temperature on the rate of water loss from a leafy shoot of a plant called *Acuba japonica*. Here are his results, showing the cumulative total for water lost.

Table 1

Time/min	Volume of water lost at 15 °C/mm³	Volume of water lost at 25 °C/ mm³
1	3.9	5.9
2	7.9	11.4
3	11.0	17.3
4	14.9	23.6
5	19.6	28.3
6	23.6	32.8
7	26.7	37.7
8	31.4	43.2
9	35.3	47.9
10	40.8	52.9

He plotted these results on graph paper, showing the rate of water loss against time (see *Fig. 5*).

Fig. 5 *Graph of Mwamba's results showing water loss from a leafy shoot at 15 °C and 25 °C*

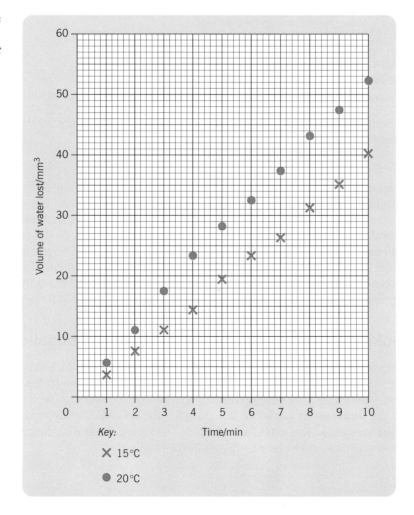

Even before plotting the curve to fit, it is clear that time is the independent variable and should be plotted on the *x*-axis and that the 15 °C and 25 °C data should be plotted as separate curves. There is a relationship between the time and the amount of water lost at each temperature. We can see that there is a **positive correlation**. As time increases, so does the amount of water lost. Although some points do not fit perfectly on a line, the data follow a definite upward trend.

On the following summary graphs, in *Fig. 6*, the letter *r* is used as the **correlation coefficient**. (The Greek letter *ρ* (pronounced 'rho') is also used.) The value of *r* ranges from +1 to −1. A value of *r* = +1 means a perfectly positive correlation, *r* = −1 indicates a perfectly negative correlation and 0 signifies nil correlation.

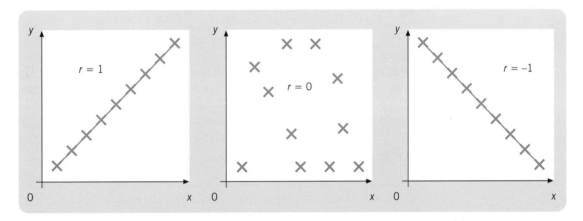

Fig. 6 *Positive, nil and negative correlation*

Obviously, with some data the lines pass through all points of either positive or negative correlations, but it is rare for biologists to obtain such 'perfect' data. How do we know where to draw our line to fit the points?

8.2 Drawing lines of best fit

It is useful to draw a simple graph of your data before doing too much analysis, so that you get an idea as to whether there is any relationship at all.

HINT ⟩ *As you collect data in an experiment, have a piece of paper to hand (it need not be a piece of graph paper) and roughly sketch out the plots as they are recorded.*

Does your scattergram show a correlation between the two variables being studied? What can be said about the relationship between the *r*-value and the appearance of the scattergram? You should notice that as the scatter spread is reduced, the correlation between *x* and *y* becomes closer and the correlation coefficient approaches 1. With an *r*-value of 0.5 or lower you can be less confident. (See *Fig. 7*.)

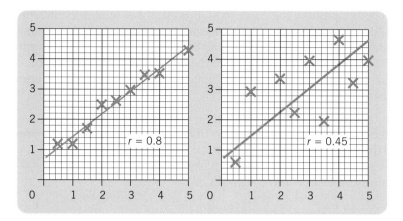

Fig. 7 *Connection between r-value and spread of scattergram*

Later we will examine analysis of data using statistical methods. Here we will just concentrate on one useful skill, that of drawing a **line of best fit** (also known as a **regression line**). At its simplest, the line of best fit could be thought of as just the neatest way to show an approximate curve. However, it really is a statistical method: taking data from a limited sample in order to estimate the relationship between x and y. As the name suggests, it is a line that represents a *tendency* (i.e. is it a positive or a negative relationship?) – but in the most accurate way (i.e. is it a strong or a weak correlation and how strong or weak?)

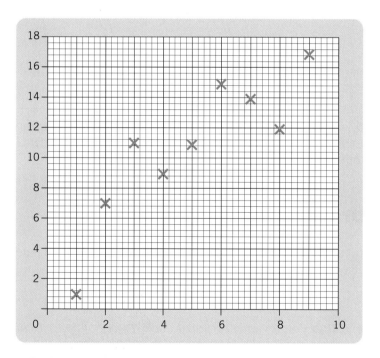

Fig. 8 *A typical scattergram*

Fig. 8 shows some data points plotted on a graph. How would you go about drawing a line of best fit between these points?

There are a number of ways you could choose, depending on the way you have collected the data.

Method 1

Use a transparent ruler; place it over the points and move it about until it looks as though the edge passes through the middle of the points, and then draw a line (see *Fig. 9*).

Fig. 9 Best fit line – using the ruler method

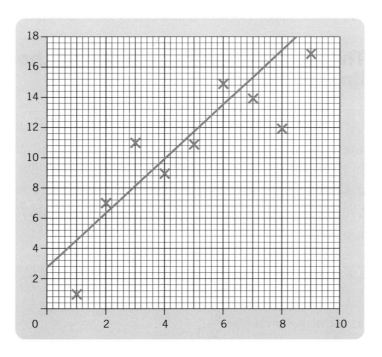

Method 2

Draw two lines on the scattergram, each one through the points furthest away from the central line and with all of the other points enclosed. Then draw a straight line equidistant between these two lines (see *Fig. 10*). That is another attempt at the line of best fit.

Fig. 10 Best fit line – using the parallel line method

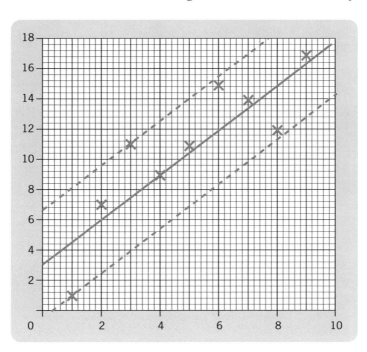

Method 3

For methods 1 and 2 you place the line 'by eye'. To be sure of one point on the line you can calculate the mean (average) of the x-values ($\bar{x}$, known as 'x bar', see p. 114) and the mean values for y ($\bar{y}$, known as 'y-bar'). These values give the co-ordinates ($\bar{x}$, $\bar{y}$) which can be plotted as a point through which the line of best fit must be drawn.

8.3 The accurate way of drawing the regression line

Methods 1, 2, and 3 will only give estimated lines of best fit. The most accurate line is defined by its two characteristics, intercept and gradient – the method of calculation follows.

● The **intercept**, which is the value of the one variable when the other is zero. So in *Fig. 11*, when $x = 0$ the line intercepts (crosses) the y-axis at a point below 5, in fact at 3.33.

● The **gradient** or slope, which can be measured by taking two points on the line and reading their values on both axes.

$$\text{Gradient} = \frac{\text{rise}}{\text{run}}$$

$$\text{or Gradient} = \frac{\text{difference in } y\text{-values}}{\text{difference in } x\text{-values}}$$

$$\left(\text{This is sometimes shown as } \frac{\Delta y}{\Delta x}.\right)$$

So in *Fig. 11*, gradient $= \dfrac{(10-5)}{(8-2)} = \dfrac{5}{6} = 0.833$

There is an equation that uses this information.

KEY FACT *The equation is $y = mx + c$, where m is the gradient and c is the intercept.*

So from the above example, we can state that

$$y = 0.833x + 3.33$$

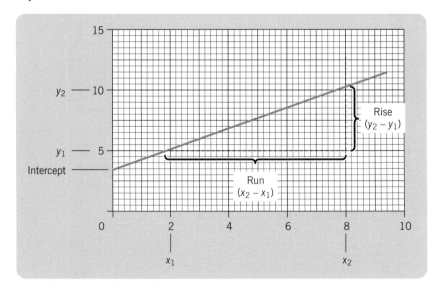

Fig. 11 Gradient calculation and intercept

Using that formula we can now draw a straight line between two points that we can calculate, using, say $x = 1$ and $x = 7$:

If $x = 1$, then
$$y = 0.833 + 3.33 = 4.16$$

If $x = 7$, then
$$y = (0.833 \times 7) + 3.33 = 9.14$$

The co-ordinates for the two points are (1, 4.2) and (7, 9.1). They can be plotted on the graph and the true, calculated, regression line can be drawn as a straight line between them.

Test Yourself

Exercise 8.2.1

1 Draw a scattergram of the following data. Draw the line of best fit. Determine a suitable linear equation for the line.

Tail length/cm, x	8.5	7.3	7.5	7.0	7.8	8.6	7.6	8.2	7.2
Body mass/g, y	29.0	13.0	22.0	16.0	19.5	16.0	20.5	22.5	16.5

Tail length/cm, x	8.0	7.5	8.1	8.5	7.6	7.8	7.5	6.8
Body mass/g, y	20.5	18.5	21.0	21.0	17.5	21.5	19.0	13.5

HINT

The value of the regression line is that you can now use it to predict values of the dependent variable for any given value of the independent variable – in other words we can be accurate in **interpolation.**

Using a scientific calculator for $y = mx + c$

If you are happy with the meanings of the terms gradient and intercept and have checked that the data from an investigation appear to be in a straight line relationship, you can use your calculator to find co-ordinates for the line of best fit. It is quite straightforward, but the calculator probably uses different symbols for the equation i.e.

$y = A + Bx$

A will give us the intercept and *B* the gradient.

The calculator instructions following use the first column of Mwamba's results for water loss at 15 °C from Table 1, p. 104.

● First prepare your calculator for linear regression analysis by pressing [mode] then [3] for regression, then [1] for linear.

● Then clear the memory by pressing [shift], then [Scl], then [=].

● Now enter the co-ordinates for each point: the x-value first followed by [,], then the y-value, then [DT] for transferring data (this is the key marked [M⁺]).

- So, it is:

  1 , 3 . 9 DT then

  2 , 7 . 9 DT then

  3 , 11 . 0 DT etc. through to

  10 , 40 . 8 DT

- All of the data is now entered, so find A by pressing shift then A then = (and you should see the answer −0.67).

- To find B: press shift then B then = (and this answer should be 4.03).

The equation for the line, therefore, is

$y = -0.67 + 4.03x$

So, we have an interesting situation to deal with. The intercept on the y-axis is a minus number, −0.67, so is below the zero. We can either extend the y-axis downwards to accommodate it, or calculate the value of the x-axis intercept for $y = 0$:

$y = A + Bx$ so, $0 = -0.67 + 4.03x$

therefore $0.67 = 4.03x$ and $x = 0.17$

You could therefore plot the graph line crossing the x-axis at 0.17.

The final stage is to use the values of A and B in the equation to find co-ordinates for another point. You could select a value for x of, say, 9.0 and in that case the value of y would be:

$-0.67 + (9 \times 4.03) = 35.6$

Use 9 and 35.6 as co-ordinates of one point and the intercept (0.17, 0) as the other; join the two points for the calculated line of best fit.

Incidentally, if after finding the values of A and B you return to your calculator and press shift then r then = , you will get the value of the correlation coefficient (r or ρ). With the data already entered it works out to be 0.999, in other words very close to +1 and indicating a very strong positive correlation.

8.4 A correlation coefficient for ecologists

But what if our data have been collected from two sites, are random and show little obvious distribution to fit a regression line? An example of this would be the frequencies of occurrence of a range of different species at two sites. The best test for assessing this type of relationship is **Spearman's rank correlation test (r_s)**. Before starting with this analysis, it is important, as with all statistical tests, to be certain that the data can be analysed in this way.

What data types can be analysed with Spearman's rank correlation coefficient?
- Data points within samples should be independent from each other.

- Ordinal scale data are most suitable (or data converted to an ordinal scale using the ranks).

- The number of paired observations should be between 10 and 30 in total.

- All individuals must be selected at random from a population, and each individual must have equal chance of being selected.

Example

Two sites were studied, and the plants in these areas counted using a quadrat to give a random sampling area. For each site, the species with the greatest number was ranked as number 1. The next was ranked as 2 and so on. In this table, the two sites contain the ranks for the different species.

Species	Site A		Site B		D	D^2
	Number	Rank A (R_A)	Number	Rank B (R_B)	($R_A - R_B$)	
Daisy (*Bellis perennis*)	11	1	12	3	−2	4
Clover (*Trifolium repens*)	9	2	15	1	1	1
Buttercup (*Ranunculus bulbosus*)	8	3	8	5	−2	4
Plantain (*Plantago maritima*)	7	4	13	2	2	4
Groundsel (*Senecio jacobaea*)	6	5	5	8	−3	9
Cat's ear (*Hypochaeris radicata*)	5	6.5	7	6	0.5	0.25
Herb Robert (*Geranium robertianum*)	5	6.5	10	4	2.5	6.25
Chickweed (*Stellaria media*)	4	8	6	7	1	1
Charlock (*Sinapsis arvensis*)	3	9	3	9	0	0
Fumitory (*Fumaria offinalis*)	1	10	2	10	0	0

HINT *Where we have equal numbers, e.g. Cat's ear and Herb Robert, we take the mean and give them both an equal rank; they occupy the 6th and 7th position, so we give each a rank of 6.5 and the next one is in the 8th position.*

(*D* is the difference between the two ranks and D^2 is the square of this difference – this is a useful trick in mathematics because the sum of the differences is zero but by squaring we get rid of negative values; sometimes the symbols *d* and d^2 are used instead.)

We can start with a **null hypothesis** (see also p. 138) that there is no significant difference between the two rank orders. Now that we have ranked the data and found the differences between the two sets, and squared this difference, we can compute the correlation coefficient by using the following formula:

KEY FACT

$$r_s = 1 - \frac{6\Sigma D^2}{n(n^2 - 1)}$$

Where D = the difference between the ranks and n = the number of pairs of values in the data.

If we then apply the data above to the formula we get:

$$\Sigma D^2 = 4 + 1 + 4 + 4 + 9 + 0.25 + 6.25 + 1 + 0 + 0 = 29.5$$
$$n = 10; n^2 = 100$$

We can then input these values into the equation:

$$r_s = 1 - \frac{6\Sigma D^2}{10(100 - 1)} = 1 - \frac{(6 \times 29.5)}{990} = 1 - 0.179 = 0.821$$

The final part in the process is to look up the Spearman's rank coefficient in a table of critical values that correspond to the number of pairs of measurements (n) in the table (see Table 1).

In this case, the critical values for r_s for 10 pairs of measurements is 0.648.

Because the magnitude (size) of the calculated value of r_s (0.821) is greater than the critical value at the 5% level ($p = 0.05$, see p. 146), our null hypothesis is incorrect and there is in fact a correlation between the sets of measurements.

Table 1 *Spearman rank correlation coefficient (table of critical values at the $p = 0.05$ level)*

n	Critical value of r_s
5	1.000
6	0.886
7	0.786
8	0.738
9	0.683
10	0.648
12	0.591
14	0.544
16	0.506
18	0.475
20	0.450
24	0.409
30	0.364

Whether or not the value for r_s is positive or negative will indicate whether or not the correlation is positive or negative; i.e. if the r_s is negative, this means that one quantity decreases linearly as the other increases. In our example there is a very strong positive correlation.

Remember **correlation** is a measure of the *strength* of the relationship between two variables. **Regression** describes the relationship between the variables and is only used if the correlation appears to be strong enough to assume that there is an underlying relationship (i.e. a linear relationship)

Exam Questions

Exam type questions to test understanding of Chapter 8

1 During the month of September two communities of sea birds were observed at feeding sites in South Wales. Site A was a large muddy beach and Site B a river channel. The 7 most frequently observed species observed on both sites are listed in the table.

Species	Site A	Site B
Curlew	150	10
Redshank	130	30
Turnstone	100	3
Oystercatcher	80	5
Greenshank	60	7
Ringed Plover	20	10
Dunlin	12	6

The null hypothesis is that there is no correlation between the distribution of numbers at feeding sites for these species. Calculate the Spearman rank correlation coefficient r_s (when $p=0.05$) and decide whether the null hypothesis can be supported.

2 Students surveying the vegetation of a saltmarsh in Anglesey compared two 10 m × 10 m sites. In one area of the saltmarsh, the mud is very salty because of the evaporation of sea water (panning), in the other area there was no panning. They recorded each plant at 200 random points in each plot.

Saltmarsh species	Site A (mud dissected by pans)	Site B (unpanned mud)
Armeria maritima	14	33
Aster tripolium	30	17
Plantago maritima	49	1
Puccinellia maritima	27	107
Salicornia sp.	9	26
Spergularia media	1	1
Triglochin maritima	69	9
Bare ground	1	6

(Field Studies, *Vol 3, No 2*)

What is a suitable null hypothesis? After calculating the value of r_s (when $p=0.05$) for these data decide if the null hypothesis can be supported.

Answers to Test Yourself Questions

Exercise 8.2.1, *p. 109*
1 Drawing a line through the points with mean co-ordinates, and ensuring an equal distribution of points either side of this line, produces a line with the equation approximating to $y = 4.5x - 15.8$.

Chapter 9

What does 'mean' mean?

After completing this chapter you should be able to:

- *calculate mean, median and mode, where relevant, from laboratory or field data*
- *sort biological data into classes*
- *describe populations and samples in terms of variability*
- *distinguish between measures of central tendency.*

9.1 Averages

TRUE OR FALSE? *The average person has rather more than the average number of legs.*

At first reading, this statement may seem pretty stupid!

Then when you think about it, you realise that the word 'average' is used carelessly. The first time it is used you are expected to think about 'the usual person', 'the normal person' or 'the man in the street'. The second time the word is used, you are expected to think about something mathematical as in:

The average of the three values: 1, 5 and 6 is $\dfrac{1+5+6}{3} = 4$

So, to go back to the 'true or false' statement: because there are some people in the population with no legs or only one leg, the arithmetic average for the whole population will be less than two (<2). So the statement is true, *in a sense*, but it warns us that we should be careful with the meaning of words.

From now on, when looking for a **measure of central tendency**, we will forget about the word average. To illustrate the three measures that we most often use in biological work, that is the mean, median and mode, we will consider a certain football team winning the 1999 European Cup Final. Some biological data can be found in the programme. For the 11 players starting the game their heights were given (in centimetres) as:

193, 173, 190, 191, 180, 180, 178, 180, 176, 178, 175

We can find one type of measure of central tendency which we will now always refer to, not as the average, but as the **arithmetic mean** of the sample and distinguish it by the symbol $\bar{x}$ (which is said '*x* bar'). So for this set of 11 players:

$$\text{Mean} = \frac{\text{sum of all the measurements}}{\text{number of measurements}}$$

$$\bar{x} = \frac{193 + 173 + 190 + 191 + \dots}{11}$$

We can calculate the mean height thus:

Let each measurement in the sample be x (i.e. $x_1, x_2, x_3 \ldots x_{11}$) then the sum of all the values is $\sum x$ (which is said 'sigma x').
In this case:

$\sum x = 1994$ cm

Let the number of individuals in the sample be n (in this case 11) then the mean:

$$(\bar{x}) = \frac{\sum x}{n} = \text{ or } \frac{1994}{11} = 181.3 \text{ cm}$$

So the mean height of these 11 players is 181.3 cm.

The next measure of central tendency is quite obvious. If we were to line up these eleven people, shortest at one end, tallest at the other, then the middle person would represent the **median**. He is literally at the *middle* of the line; there are as many players shorter than him as there are taller than him.

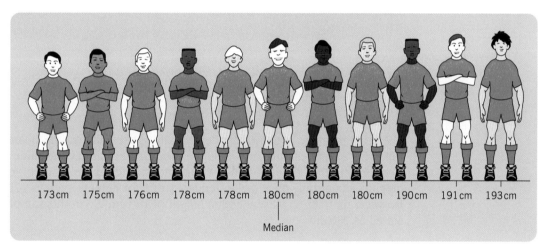

173 cm 175 cm 176 cm 178 cm 178 cm 180 cm 180 cm 180 cm 190 cm 191 cm 193 cm

Median

Fig. 1 *The median of the football team*

173 175 176 178 178 180 180 180 190 191 193
↑
The median height

So the median height in this team is 180 cm.

What would happen after the first substitution when a player 185 cm tall came on to the field?

Now we have to find the median of a sample of 12 players who took part in the game – there is no 'middle of the line'. So we have to take the 6th and 7th and find the mean of these two heights.

173 175 176 178 178 180 180 180 185 190 191 193
↑
The median height

In this case, there is no change because the new median is:

$$\frac{180 + 180}{2} = 180$$

However the arithmetic mean has changed. By how much?

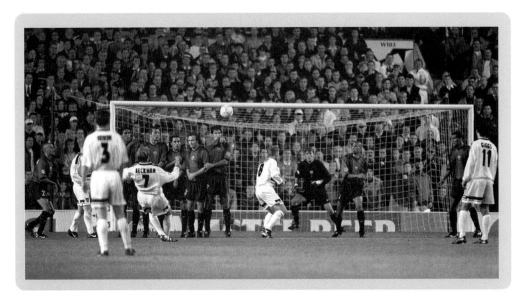

Fig. 2 *The scoring moment by one of the 180 cm Manchester United Players (see p. 115)*

The other measure of central tendency that we use will become clearer with some of the following examples. It is known as the **mode** and is simply the value that occurs most frequently in the population that we are measuring. In the group of footballers, most of the heights appear only once (i.e. a frequency of 1 for heights such as 173 cm and 191 cm). However there are 3 players with a height of 180 cm. So in this sample, the mode can be stated as 180 cm.

Test Yourself

Exercise 9.1.1

On a sunny afternoon, with nothing else to do, Megan was pulling 'petals' (actually, they are ray florets!) from daisies. She had thought that all daisies were the same, but was surprised to find considerable variation in the number of florets. She counted 25 daisies and recorded the score for each one. She then made a tally chart, having first decided to group them into number classes e.g. 20–4 florets would be one class.

Field notes

44	39	46	33	38
44	36	51	31	28
27	29	42	41	33
32	36	47	35	36
35	37	24	33	22

Number class	Tally chart
20–4	⊞
25–9	⊞
30–4	IIII
35–9	IIII III
40–4	IIII
45–9	II
50–5	I

Fig. 3 Daisy

1 What is the mean number of ray florets in the sample?

2 How would you find the median?

3 What is the mode?

Test Yourself

Exercise 9.1.2

1 Find $\bar{x}$ for the age in months of 6 friends.

2 Use your calculator to find $\bar{x}$ for the number of ray florets in the field notes (on the previous p. 116.)

You will be using this technique again in the next chapter

HINT

Calculate the mean of 174, 85, 63 and 125.

Find Σ, then divide by the number of measurements:

$\boxed{1}\boxed{7}\boxed{4}\boxed{+}\boxed{8}\boxed{5}\boxed{+}\boxed{6}\boxed{3}\boxed{+}\boxed{1}\boxed{2}\boxed{5}\boxed{=}\boxed{\div}\boxed{4}\boxed{=}$ *answer*

If you are in-putting a great deal of data – as you would in a field investigation – then you should be familiar with the keys to use on the scientific calculator.

So to get into SD mode (standard deviation, see p. 123):

$\boxed{mode}\boxed{2}$

Then to clear the statistics memory: $\boxed{shift}\boxed{Scl}\boxed{=}$
Then you input the values, following each one with $\boxed{DT}$ *thus:*

$\boxed{1}\boxed{7}\boxed{4}\boxed{DT}\boxed{8}\boxed{5}\boxed{DT}\boxed{6}\boxed{3}\boxed{DT}\boxed{1}\boxed{2}\boxed{5}\boxed{DT}\boxed{\div}\boxed{4}\boxed{=}$ *answer*

To find $\bar{x}$ (x-bar): $\boxed{shift}\boxed{\bar{x}}\boxed{=}$

The answer should now be displayed as 111.75.

9.2 Describing populations and samples

The observations that Megan made in Exercise 9.1.1 are said to be **discrete**, in other words only certain numbers of ray florets are possible – the *whole* numbers. It would be the same for the number of different species of lichen on a tombstone or of species of birds living in a hedge. (You wouldn't expect to find 0.7 of a species!) If we make observations of the heights of our colleagues, we are only limited by the accuracy of our technique – any number could be possible, a person could be 1.762 m tall. Observations of this type are **continuous**. The third type of observation is **categorical** (or nominal). This is where there are a limited number of categories which do not fit into a number scale. It could be the categories of food in the diet (simply lipid, carbohydrate and protein), blood groups, names of the species in an angler's catch or environmental characteristics (sunny or shaded). The **frequency** is the number of individuals recorded in any of these groups.

All scientists expect some **variability** in results and take pains to iron out errors that could be caused by their inaccurate measurements. Biologists have particular problems in this respect, because biological structures vary greatly. You could observe a small grove of 20-year old oak trees. Although they were grown from acorns from the same parent tree, planted at the same time, in the same soil with the same climate, there is likely to be variability in height. You can probably suggest a number of genetic and environmental reasons for this. It is worth noting that in biological investigations, there is a greater chance of variability due to the range of the material than to the inaccuracy of the experimenter!

It means that we must work with a *large* sample (so we must be sure to collect enough data for our sample, i.e. we must have a large enough sample properly collected to represent the whole population). We must also measure as accurately as possible and even carry out replicate surveys. Perhaps with the 100 oak trees you could just about measure each one. If we were to study a huge population (say the resistance to tsetse fly in the wildebeest of the Serengeti) we would have to investigate a small random sample, and if it is collected properly it will represent the whole population.

Later we will examine how data can be treated to see if the sample is typical of the population.

Frequency distributions

The table below gives the details for a herd of 94 cows, showing the milk yield per cow over a year.

Category	Milk yield per cow/dm^3	Number of cows (frequency)
1	2000–499	1
2	2500–999	3
3	3000–499	5
4	3500–999	7
5	4000–499	13
6	4500–999	18
7	5000–499	17
8	5500–999	13
9	6000–499	8
10	6500–999	5
11	7000–499	3
12	7500–999	1

Note that the herd has been classified into 12 categories of milk yield. Each cow in the herd can be placed into one of the categories and we can state the **frequency** for each yield group:

In category 7 (5000 – 499 dm³) there are 17 cows, so the frequency is 17.

This allows us to draw graphs with categories on the *x*-axis and frequencies on the *y*-axis (see *Fig. 4*).

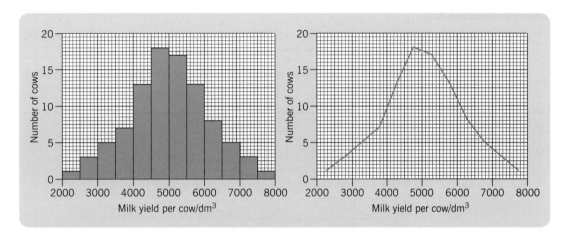

Fig. 4 *Two types of frequency distribution of milk yield*

Can you work out the modal class for milk yield for this herd?

Yes, it's category 6.

It is not possible, however, to calculate an accurate mean or median value from the data. However, the graph does give some information:

- 16 cows have a low yield and may not be worth the cost of keep in terms of their productivity.

- Some cows could certainly be used as good breeding stock for improving the average yield of future generations of the herd.

Note that the graphs have an interesting shape. The majority of cows have a yield near to the mean value and there is a decline in frequency in the categories on either side of the mean. This type of distribution is a **normal distribution** (see p. 124 for more precise details). It is possible to draw a curve to smooth it out, it can then be described as a bell-shaped curve (see *Fig. 5a*).

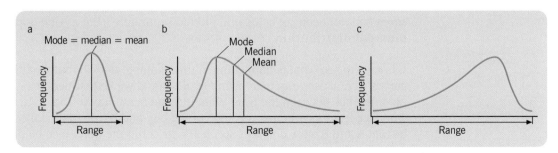

Fig. 5a *Bell-shaped curve*   **Fig. 5b** *Positively skewed distribution*   **Fig. 5c** *Negatively skewed distribution*

There are some subtle variations in the shapes of distribution curve (*see Fig. 5*).

The curve in *Fig. 5b* rises steeply to the maximum and then tails off gradually to the right. It is not symmetrical – we say that it is **positively skewed** (the mean > the median). *Fig. 5c* shows exactly the same range and the same frequency for the mode value, but the curve is skewed to the right and is said to be **negatively skewed** (the mean < the median)

Test Yourself   Exercise 9.2.1

1 Can you describe how graphs b and c in *Fig. 5*, above, could show the distribution of biology marks in two different classes?

2 In *Fig. 5b* the mean median and mode have been indicated. Mark these values on a copy of *Fig. 5c*.

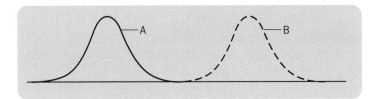

***Fig. 6** Bimodal distribution – non overlapping ranges*

When data plotted appears like *Fig. 6*, it is obvious that we have plotted values for two different populations. There is no overlap in the ranges of A and B. A could be a population of 7-year-old children and B could be of 12-year-olds (or the graph could have been of a population of short jockeys and tall guardsmen).

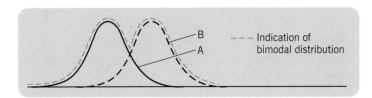

***Fig. 7** Bimodal – overlapping ranges*

The distribution in *Fig. 7* shows the heights of an adult population on an island. The females (A) and the males (B) have been plotted separately. You can easily see that the modes and the means are well separated, but the ranges overlap i.e. the tallest females are taller than the shortest males. The green curve shows what would have been plotted if the data on the two sexes had not been collected and plotted separately – but as one population. It shows a **bimodal distribution** i.e. two modes.

It is easy to see what is happening in this example (*Fig. 7*); but what if the two distribution curves were closer together, as in *Fig. 8*? Would we be justified in saying that there really was no difference between the two? What if A had been drawn from data of the body length of woodlice from a garden at the edge of the city and B from data collected 5 km away? Could it be that A and B were different species? – or could the difference be due to an environmental difference? – or is it pure chance that there appears to be a slight difference which is not really significant?

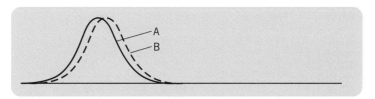

Fig. 8 *Overlapping ranges and modes very close*

To be able to answer we need to apply some statistical tests (see p. 143).

Exam Questions

Exam type questions to test understanding of Chapter 9

1 The data in the table below, came from a Scottish medical journal, published in 1817. Medical orderlies had recorded the chest measurements from a sample of over 5600 soldiers. Use the 16 length categories to plot a histogram. What is the shape of the distribution of chest circumference of these men?

Chest/in	Number of men	Chest/in	Number of men
33	5	41	925
34	20	42	650
35	80	43	315
36	190	44	150
37	400	45	50
38	750	46	20
39	1050	47	5
40	1080	48	1

2 *Ascophyllum nodosum*, the egg wrack is a seaweed, commonly found growing on inter-tidal rocks around the coast of Britain. A student measured the amount of growth in one year of 164 specimens. Her table of results is shown below arranged in 10 mm categories.
   (a) Plot a histogram of her results.
   (b) State which category is the mode.
   (c) Explain how to find the median.

Fig. 9 *Ascophyllum nodosum*

Growth/mm	Frequency	Growth/mm	Frequency
30–9	1	120–9	16
40–9	1	130–9	14
50–9	7	140–9	8
60–9	7	150–9	9
70–9	10	160–9	3
80–9	14	170–9	8
90–9	18	180–9	2
100–9	23	190–9	0
110–19	22	200–10	1

Answers to Test Yourself Questions

Exercise 9.1.1, *p. 116*
1 $\bar{x} = 35.96$ (so a sensible answer would be 36)
2 By arranging in increasing order and taking the 13th out of 25, i.e. the 13th number in the series is 36.
3 The category with the greatest number of individuals, i.e. 35–9

Exercise 9.2.1, *p. 117*
1 Well, one class could have exceptionally good students and/or very fine teaching!

Chapter 10

Distribution

After completing this chapter you should be able to:

- *describe normal distribution*
- *understand standard deviation as the spread of data about a mean*
- *calculate the standard deviation and distribution around the mean if the population is normally distributed*
- *calculate confidence limits and standard errors.*

10.1 Standard deviation

In the previous chapter we saw how there is much variation in natural populations; we examined how this variation can be plotted in a graph against the frequency of occurrence to give, in the case of the herd of cows on p. 118 a normal distribution for milk yield.

Example

88 apples from one tree were weighed and a tally chart made of the number of apples in class intervals according to their mass:

Class interval, mass/g	Number of apples in class (frequency)	Tally score
80–90	1	/
90–100	3	///
100–10	11	JHT JHT /
110–20	14	JHT JHT ////
120–30	30	JHT JHT JHT JHT JHT JHT
130–40	16	JHT JHT JHT /
140–50	9	JHT ////
150–60	2	//
160–70	2	//

If you draw a smooth line around the tally scores, you at once get an indication that the distribution seems to be normal.

HINT *In the table, there is one group of apples weighing 90–100 g and another weighing 100–10 g. You may wonder what you would do about an apple weighing exactly 100 g. You couldn't put it in both classes – the accepted way is to put it in the lower class. In other words, the class is actually '90 up to 100'; the next class is used for any value over 100 g – even 100.01 g.*

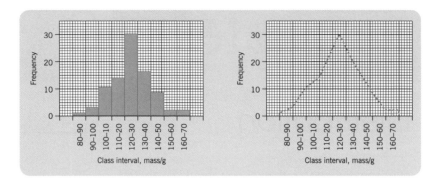

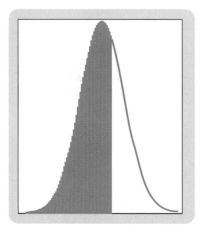

Fig. 1 Graphs showing frequencies of number of apples of different mass intervals

When the histogram is plotted, each class interval is shown as a column of the same width on the *x*-axis. The *height* of the column shows the number of apples in that class and the *area* of each of the columns is proportional to the total mass of apples in that class. As there are 30 apples between 120 g and 130 g, you could guess at a mean mass of 125 g. The total mass of apples of that size would be

30 × 125 g = 3750 g

Here the classes closest to the mean contain the most data (i.e. most apples). When the distribution curve of continuous data is plotted as a smooth curve, the **normal distribution curve** appears. Another name used for this type of distribution is the **Gaussian distribution**. It is always bell-shaped, always symmetrical; the mean, mode and median are all in the centre – just as in *Fig. 5a*, p. 119.

Perhaps a leap from the histogram to the smooth bell-shaped curve leaves a little too much to the imagination. However, if you think about a huge orchard with a crop of tens of thousands of apples, and if the masses were graphed with class intervals of only 1 g, you could imagine that it may look like the left half of *Fig. 2*.

Fig. 2

The normal distribution – more precise details

We now have to consider a *measure* of how the data are distributed about the mean. In *Fig. 3* there are three bell-shaped curves. They all have the same mean, but in A the data are distributed *widest* from the mean and in C the data are clumped *closest* to the mean.

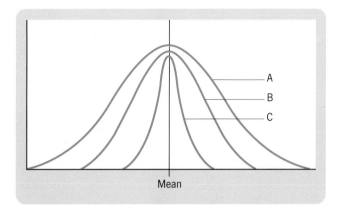

Fig. 3

The precise measure of this spread about the mean is the **standard deviation** from the mean. (*If you have not met this term before, the meaning will become clear as you continue.*) So, in *Fig. 3*, C has a *small* standard deviation and A has a *large* standard deviation. In what way can we show this?

Example
The owner of a trout farm may be giving purchasers as much information as they need when advertising, 'Our trout average 400 g and most are within 25 g of this weight'. The scientist may forgive the use of 'average' and 'weight', but would want rather more precise information. If we were told that the mean mass was 400 g with a standard deviation of 25 g it becomes clearer.

KEY FACT   *By definition, a range of values between one standard deviation above the mean and one standard deviation below the mean (i.e. a class of 375–425 g) would contain 68% of all of the trout in that population. (For the moment you don't need to know why – just accept it!)*

One of the properties of a normally distributed population is that about 95% of the population is included in a class that spreads two standard deviations above and two standard deviations below the mean.

Table 1 *Percentages of the population included in* perfect *normal distribution (σ is the symbol for standard deviation)*

Standard deviation	Percentage of population included in range			Percentage of population excluded	
	Total	Above mean	Below mean	Above mean	Below mean
1σ	68	34	34	16	16
2σ	95	47.5	47.5	2.5	2.5
3σ	99.5	49.75	49.75	0.25	0.25

The Greek letter σ is one version of the small (lower case) letter Greek sigma (Σ). The values in Table 1 can be shown on the normal Gaussian distribution curve (see *Fig. 4*).

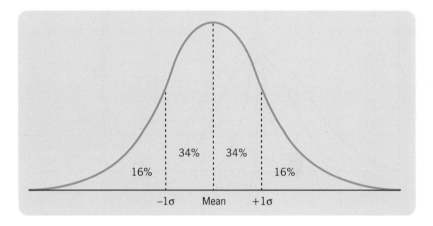

Fig. 4 *Perfect normal distribution curve*

HINT *We continue to use the symbol x̄ for mean values collected from samples. There is another symbol that you will also see – the Greek letter mu, μ. This comes from the statistical model assuming a normal distribution and is used in the exercise below.*

Test Yourself   Exercise 10.1.1

1 The fish farmer sells 10 000 trout in a year (mean mass $\mu = 400$ g and $\sigma = 25$ g). Assuming a normal distribution, estimate: the number of these that would be in the range $\mu \pm 1\sigma$.

2 The number that would have a mass greater than $\mu + 1\sigma$.

3 The number that would be lighter than $\mu - 3\sigma$.

4 The pulse rates of 2400 patients were recorded and it was calculated that the mean value was 74 beats per minute with a standard deviation of 6 beats per minute.

   What percentage of the patients had a pulse rate in the range 68–80 beats per minute? How many patients was this?

5 How many patients had a pulse rate of more than two standard deviations above the mean?

Standard deviation and samples

In the previous examples, you were given the values of the standard deviation. Now imagine you have collected your own data from a field work exercise. You have drawn a quick tally chart, with about 10 class intervals, and it seems as if it is a normal distribution. You can easily find the mean but you also need to find the standard deviation. The formula for the whole population will be given in exam questions.

KEY FACT

$$\sigma = \sqrt{\frac{\sum x^2 - \frac{(\sum x)^2}{n}}{n}}$$

Where σ = *standard deviation of the population*
x = *any one value and x^2 = the value squared so*
$\sum x^2$ = *the sum of all the values squared*
$\sum x$ = *the sum of all of the values*
$(\sum x)^2$ = *the square of the number obtained after all the values have been summed and*
n = *the number of values in the population.*

Example

x	x^2
1	1
2	4
3	9
4	16
$\sum x = 10$	$\sum x^2 = 30$
$(\sum x)^2 = 100$	

However, the equation that you will use is likely to be a slight modification of this because you will usually be looking at a *sample* standard deviation rather than at the whole population.

KEY FACT   *The sample equation to use is:*

$$\sigma_{n-1} = \sqrt{\frac{\sum x^2 - \frac{(\sum x)^2}{n}}{n-1}}$$

Where σ_{n-1} = *standard deviation of the sample,*
x = *any one value and x^2 = the value squared so*
$\sum x^2$ = *the sum of all of the values squared*
$\sum x$ = *the sum of all of the values so*
$(\sum x)^2$ = *the square of the number obtained after all values have been summed and*
n = *the number of values in the sample.*

Example
Adrian and Bushra were investigating a population of cuckoopint, *Arum maculatum*, in a woodland in Somerset. In August, the bright orange fruits are carried on a stalk above the

leaf litter. They measured the length of stalks for a sample of 20 plants and recorded (in centimetres) the following:

21.2 23.1 24.6 25.4 26.2 26.8 27.7 28.1 28.6 28.9
29.3 29.7 30.2 30.5 31.3 31.8 32.6 33.3 33.4 37.2

Fig. 5 *Fruiting stalks of **Arum maculatum** (cuckoopint)*

In their written report they had to carry out a statistical analysis to show the distribution of these results. They decided to make a table of two columns, listing in the first column all 20 values of *x* (the length of the stalks). They put them in increasing numerical order – but before going any further they made a tally chart to see if the data had come from a normal distribution.

Table 2

Class intervals, length of stalk/cm	Frequency	Tally score
20–21.99	1	*I*
22–23.99	1	*I*
24–25.99	2	*II*
26–27.99	3	*III*
28–29.99	5	*Ʃℋ*
30–31.99	4	*IIII*
32–33.99	3	*III*
34–35.99	0	
36–37.99	1	*I*

They decided that the tally chart looked as if the data had come from a normal distribution, so they proceeded to add all of the values of *x* to get $\sum x$.

Table 3

x/cm	x^2
21.2	449.44
23.1	533.61
24.6	605.16
25.4	645.16
26.2	686.44
26.8	718.24
27.7	767.29
28.1	789.61
28.6	817.96
28.9	835.21
29.3	858.49
29.7	882.09
30.2	912.04
30.5	930.25
31.3	979.69
31.8	1011.24
32.6	1062.76
33.3	1108.89
33.4	1115.56
37.2	1383.84
$\Sigma x = 579.9$	$\Sigma x^2 = 17\,092.97$

So, as n was 20, $n - 1 = 19$. These values can now be substituted into the equation:

$$\sigma_{n-1} = \sqrt{\dfrac{17\,092.97 - \dfrac{336\,284.01}{20}}{19}}$$

Standard deviation of the sample,

So, $\sigma_{n-1} = \sqrt{14.672} = 3.83$

They were able to state in their report that the mean length was:

$$\dfrac{579.9 \text{ cm}}{20} = 29.0 \text{ cm}$$

and the sample standard deviation was 3.8 cm (both given to 1 decimal place, as the original measurements).

Try to remember these stages for calculating sample standard deviation:

Stage 1   List data.

Stage 2   Make rough tally chart to check data is normally distributed.

Stage 3   Add all measurements to find $\sum x$.

Stage 4   Square each value of x to give x^2.

Stage 5   Add all x^2 values to give $\sum x^2$.

Stage 6   Find $n - 1$.

Stage 7   Substitute into equation.

HINT *Your scientific calculator will work this out for you, by the means of a key labelled* $\boxed{x\sigma_{n-1}}$.

The method starts the same way as the method for calculating the mean, shown on p. 117

- *Change to standard deviation mode:* $\boxed{\text{mode}}$ $\boxed{2}$
- *Clear statistics memory:* $\boxed{\text{shift}}$ $\boxed{\text{Scl}}$ $\boxed{=}$
- *Input the values of* x, *each followed by* $\boxed{\text{DT}}$. *So from Table 3:*

  $\boxed{2}$ $\boxed{1}$ $\boxed{.}$ $\boxed{2}$ $\boxed{\text{DT}}$ $\boxed{2}$ $\boxed{3}$ $\boxed{.}$ $\boxed{1}$ $\boxed{\text{DT}}$ ... $\boxed{3}$ $\boxed{7}$ $\boxed{.}$ $\boxed{2}$ $\boxed{\text{DT}}$

- *Then to find standard deviation:* $\boxed{\text{shift}}$ $\boxed{x\sigma_{n-1}}$ $\boxed{=}$

The value should be displayed as 3.8.

This obviously takes much *less time!*

Test Yourself   Exercise 10.1.2

Use your calculator to go through the stages of the cuckoopint calculation pp. 128–9.

1 What would you expect the length range to be for 95% of the sample?

2 What percentage of the population would be longer than 32.8 cm?

WARNING *The value that we have calculated here is for the sample. We have used the symbol* σ_{n-1} *but some calculators, books and examination papers may use the symbol s or even* s_{n-1}.

If working out the standard deviation of the whole population (but you probably won't have to do that!) you may find the symbols used are σ, σ_n, $x\sigma_n$ *or* s_n.

HINT *If you have collected data that the tally chart indicates is not normally distributed (e.g. it could be very skewed or evenly spread), then there is no point in attempting to describe the distribution around the mean. Don't waste time with standard deviations. It is better to state the* median *only, so as to give a central measure.*

10.2 Confidence

HINT

NB You may not have to know about this for the exam that you are taking. It is included because you may want to use the technique as part of your individual investigation.

Look back to the data for *Arum maculatum* (p. 129). The calculation showed that the sample mean was 29.0 cm and the sample standard deviation was 3.8 cm. If another sample was collected from the same area, the values could be, say, 28.3 cm and 3.6 cm.

How confident would an ecologist be that either of these values for the samples were typical of the whole population? Could he/she be 100% confident or perhaps only 50%? Biologists generally work from a 95% confidence level (though sometimes, as in medical studies, it is necessary to be 99.9% confident).

As you probably expected there is an equation for working it out. It is:

KEY FACT

$$95\% \text{ confidence limit} = \bar{x} \pm t \times \frac{s}{\sqrt{n}}$$

$\bar{x}$ is the mean of the sample and $t = 1.96$ (where $n > 30$).
There are tables such as this to consult; you should really only use the test if there are more than 30 in the sample (though for exam projects numbers down to 10 can be calculated from the table).

n	t
10	2.26
15	2.15
20	2.09
25	2.00
30+	1.96

s = standard deviation for the sample
So, for the *A. maculatum* sample:

$$95\% \text{ confidence limit} = 29.0 \text{ cm} \pm 2.09 \times \frac{3.8}{\sqrt{20}}$$

$$= 29.0 \text{ cm} \pm 1.78 \text{ cm or } 1.8 \text{ cm (to 1 d.p.)}$$

This means that we can be 95% confident that the mean of all of the stalks in the whole population will be within these limits. In other words, there is only a 5% chance that the actual mean is outside this range.

KEY FACT

$\frac{s}{\sqrt{n}}$ *is known as the* **standard error**.

HINT

Error bars can be drawn using confidence limits or standard errors, or even standard deviations (as in Fig. 12, p. 98). It doesn't matter which you draw, as long as you state what you are showing.

Exam Questions

Exam type questions to test understanding of Chapter 10

HINT › *In all these questions you should assume a normal distribution.*

1 If the mean value of IQ in the British population is 100 with a standard deviation of 15:

   (a) Between what IQs do 95% of the population fall?
   (b) What proportion of the population have an IQ of over 130?

2 What can you say about a population of insects with a mean wing length of 38.3 mm and a standard deviation of 2.7 mm?

3 A group of people have a mean height of 174.0 cm and a standard deviation of 6.5 cm. What are the upper and lower extremes of height that fall

   (a) within 2 standard deviations of the mean?
   (b) within 3 standard deviations of the mean?

4 A marine ecologist working on a rocky shore in Devon measured the heights of 19 dog whelks (*Nucella lapillus*) and recorded in millimetres:

   28 30 31 31 33 33 33 36 36 36 36 36 39 39 39 40 40 42 44

   Calculate the mean and standard deviation.

5

Surface area of leaves/cm^2	
Deep shade (site A)	Open clearing (site B)
21	15
14	17
16	18
18	17
19	17
21	19
19	13
22	14
18	21
16	13
13	16
22	13
21	16
23	12
19	14
18	12
15	20
Mean $\bar{x}_A$ =	Mean $\bar{x}_B$ =
Standard deviation of A =	Standard deviation of B =

Leaves which are adapted to low light intensities are known as *shade leaves*, while those which function more efficiently in high light intensities are known as *sun leaves*. An investigation was carried out to determine whether there was a significant difference in the surface area of shade and sun leaves of dog's mercury (*Mercurialis perennis*), a plant which grows in woodland and shady places. 17 leaves of dog's mercury were collected from plants growing in deep shade (site A) and 17 leaves from plants growing in a clearing open to sunlight (site B). The surface area of each leaf was measured. The results are shown in the table.

Calculate the mean and standard deviation for each collection (site A and site B).

(Adapted from Edexcel, HB6, June 1997)

Answers to Test Yourself Questions

Exercise 10.1.1, *p. 126*
1 (68% of 10 000) = 6800
2 (16% of 10 000) = 1600
3 (0.25% of 10 000) = 25
4 (68% of 2400) = 1632 patients
5 (2.5% of 2400) = 60 patients

Exercise 10.1.2, *p. 130*
1 21.4–36.6 cm
2 16% (above one standard deviation from the mean)

Chapter 11

Chi-squared: a test of closeness

After completing this chapter you should be able to:

- *state a null hypothesis*
- *calculate the value of χ^2 from given data for a given probability value*
- *use the χ^2 test for data from genetics and ecology*
- *use 2 × 2 contingency tables.*

11.1 Chi-squared test and the null hypothesis

Mendel again! – we have already looked at some of Mendel's actual results (p. 53 and p. 55) and stated that a ratio that he obtained of 2.84 : 1 could be rounded up to 3 : 1. Then when dealing with a dihybrid ratio, we assumed that 9.84 : 3.16 : 3.38 : 1 could be considered to be a 9 : 3 : 3 : 1 ratio. You may have thought that 9.84 is much nearer to 10 – so why isn't it a 10 : 3 : 3 : 1 ratio?

Well, genetics theory tells us that the former is to be *expected* (remember the calculation on p. 56) and there is a neat statistical test that tells us how close a set of actual results is to a predicted (or expected) result. It is known as the χ^2 test – said as 'chi-squared' (another Greek letter, χ, which is pronounced as 'kai' and rhymes with 'eye'). To carry out the test using some of Mendel's records, we need to work out the value of χ^2 and then check it in a table of values.

Example

This example uses the values from Table 1 (p. 53) for length of stem. In the total of 1064 pea plants, 787 were tall and 277 were dwarf. These are the **observed values** (*O*). What if the 1064 were divided *exactly* into $\frac{3}{4}$ tall and $\frac{1}{4}$ dwarf? The **expected value**: (*E*) would be

$$1064 \times \tfrac{3}{4} = 798 \text{ for tall and } 1064 \times \tfrac{1}{4} = 266 \text{ for dwarf}$$

We can now make a table to show these values and the next stages of the calculation.

	Observed (*O*)	Expected (*E*)	*O* – *E*	(*O* – *E*)^2	$\dfrac{(O-E)^2}{E}$
Tall pea plants	787	798	–11	121	$\dfrac{121}{798} = 0.15$
Dwarf pea plants	277	266	+11	121	$\dfrac{121}{266} = 0.45$

> **HINT** *Actually, we can ignore the + and – sign in the O – E column, because we are squaring this difference in the next column.*

Add together the numbers in the last column:

$0.15 + 0.45 = 0.60$ This gives the value χ^2.

KEY FACT *The chi-squared equation that sums up this sequence of operations is:*

$$\chi^2 = \sum \frac{(O - E)^2}{E} = 0.60 \ (\text{for this example})$$

We now need to look up the value of chi-squared in a printed table of critical values. This will allow us to judge if there is a significant difference between the observed and expected values. This is part of a chi-squared table. A full table is given on p. 153.

Table 1 *Part of a χ^2 table*

Degrees of freedom	Level of significance (probability, p)	
	$p = 0.05$ (or 5%)	$p = 0.01$ (or 1%)
1	3.84	6.63
2	5.99	9.21
3	7.81	11.34
4	9.49	13.28
...	...	...
12	21.03	26.22
...	...	...
20	31.41	37.57

You first need to know how many **degrees of freedom** are involved. You don't need to know the meaning of this term – just how its value is calculated (as number of categories minus one). In our example there are only two categories, tall and dwarf, so $2 - 1$ gives one degree of freedom. Table 1 gives a critical value for χ^2 with one degree of freedom of 3.84 (at the $p = 0.05$ level).

HINT *The $p = 0.05$ level is the one that you will be using unless told otherwise. It is the level most used in investigations and is explained on p. 146.*

Well, what can we do with the value of 3.84 that we got from the table? Our calculated value of 0.60 is *well below* this critical value; the rule is that this means there is no significant difference, at the $p = 0.05$ level, between the observed and expected results. So it is fine to accept that a tall to dwarf ratio of $3 : 1$ is an acceptable statement to make here.

Test Yourself

1 (a) Use the method explained above to calculate the value of χ^2 for another set of Mendel's observations (p. 55) where we suggested that $315 : 101 : 108 : 32$ could be accepted as a $9 : 3 : 3 : 1$ ratio.

> **HINT** *Remember after calculating your table that, this time, there are four categories to be added to give χ^2.*

(b) Now check in the χ^2 table whether or not, at the $p = 0.05$ level, the $9 : 3 : 3 : 1$ ratio can be accepted with Mendel's observed numbers.

> **IMPORTANT** *The chi-squared test can only be used if the data are **categorical** and based on counts not measures. We can use the categories tall/dwarf because these are the two possible groups in peas. We couldn't use height in humans because those data would be **continuous**.*

The chi-squared test is ideal for genetics data, because we expect from theory certain ratios for specific categories. It can also be used for the results of experiments in animal behaviour. We could test if woodlice selected dry or humid conditions in a choice chamber, because we would be using two categories – dry or humid.

Example

The following is an example of a simple ecological investigation.

The butterfly bush (*Buddleia davidii*) is a common garden shrub visited frequently, as the name suggests, by butterflies collecting nectar.

In one period of observation, 46 visits were recorded to a bush with *purple* flowers. In the same time, there were 28 visits to a bush with *white* flowers. The question to be asked here is: 'is this a significant difference, or are the numbers close enough for us to accept that there is no significant preference of colours?'.

Scientists have to be suspicious of any change from the usual, so the standard scientific method is to assume that there is *no* preference for either colour. The way to start an investigation is to develop a **null hypothesis**. In this case we would say that 'butterflies are *equally* likely to visit white and purple *Buddleia* flowers'.

That would mean that for the chi-squared test, the expected numbers (*E*) would be *equal* for white and purple as shown in the table below (i.e. the total of the 74 visits would be equally divided – 37 to each colour).

	O	E	$(O - E)$	$(O - E)^2$	$\dfrac{(O - E)^2}{E}$
Visits to purple flowers	46	37	+9	81	$\dfrac{81}{37}$
Visits to white flowers	28	37	−9	81	$\dfrac{81}{37}$

$$\chi^2 = \frac{81}{37} + \frac{81}{37} = 4.38$$

As there are two categories, the number of degrees of freedom (2 – 1) is again 1. By looking back to the first row of the χ^2 table, p. 135, we can see that our calculated value of 4.38 is *above* the value in the table of 3.84.

This means that we must *reject* the null hypothesis. So there *is* a significant difference in the frequency of the visits to the different coloured flowers.

KEY FACT   *If χ^2 is above the critical value – reject the null hypothesis.*
If χ^2 is below the critical value – accept the null hypothesis. (Actually, it is safer to say that there is not strong enough evidence to reject it.)

Example

Recently a drug trial was carried out on a new anti-malarial treatment. As in many such investigations, the medical biologists needed an answer to the question: 'Is the effect on Group A different from the effect on Group B?'. They carried out a double-blind test (neither the subjects nor the experimenters knew if the numbered box of white tablets contained the old drug or the new one; the key to the numbers was locked away in the safe of an independent person). 400 people were subjected to each treatment. All 200 volunteers in Group A received the new drug and the 200 in group B were given the older drug.

In the usual way, the investigation needed to start with the null hypothesis: 'there is no difference in the effects of the treatment received by Group A and Group B'.
In the case of this trial of relative efficacy, the statistical tests showed that the null hypothesis should be rejected. The new treatment was more effective and had fewer side effects. In the year 2001 it became a prescribable drug.

Test Yourself   Exercise 11.1.2

1 In a drug trial of a new antibiotic A, its effects were compared with the effects of an antibiotic B – already in use. Two equal-sized groups were tested. Antibiotic A was given to all of those in Group A and antibiotic B to the people in Group B. At the end of 6 days, everyone was assessed. It was found that the wounds of 14 of those in Group A had healed cleanly; this result was also seen in 24 of Group B. Does a chi-squared test show that this is a significant difference?

HINT ▷   *State first the null hypothesis. How many categories are there? You are given the values for O – what are the values for E?*

KEY FACT   Remember *that for chi-squared tests:*

- *We can only use categories – not continuous measurements such as length or mass.*
- *It is important to state the probability.*
- *The test is only effective with reasonably large numbers – the rule is that each 'expected value' should be more than 5.*
- *We are looking for a measure of agreement between observed and expected results.*

11.2 Contingency tables

The examples looked at so far give one set of observations in 2 or 4 categories. Now examine the data from a different type of investigation.

Fig. 1 *Ecology students surveying a shingle beach at Slapton, Devon*

Example

Three students were looking at the vegetation on a shingle beach. They were interested in two species: sea campion (*Silene maritima*) and rest-harrow (*Ononis repens*). They wondered if there was a relationship between the two species, i.e. if they were found growing together in a 'linked way' or purely by chance. In other words, were they independent of each other? They looked at the ground within 160 random quadrats and recorded their observations in 4 categories.

Table 2 *Observed frequencies (O)*

		Ononis repens	
		present	absent
Silene maritima	present	64	16
	absent	34	46

They decided to set up a null hypothesis (sometimes known as H_0) that there is *no* link between them. The alternative hypothesis (H_1), would be that there *is* an association between the presence of the two species. So the next task is to calculate the expected frequencies (*E*) based on independence. This is done for each box in the table, using the formula:

$$E = \frac{\text{column total}}{\text{grand total}} \times \text{row total}$$

Table 2 *Calculating expected frequencies (E)*

		Ononis repens		Row totals
		present	absent	
Silene maritima	present	*a* 64	*b* 16	80
	absent	*c* 34	*d* 46	80
	Column totals	98	62	Grand total = 160

The expected frequencies for boxes a, b, c and d in Table 2 are:

$$a = \frac{98}{160} \times 80 = 49; \quad b = \frac{62}{160} \times 80 = 31; \quad c = \frac{98}{160} \times 80 = 49; \quad d = \frac{62}{160} \times 80 = 31$$

Now for each box calculate $\frac{(O-E)^2}{E}$ and add together the 4 values, i.e. for box *a*

$$\frac{(64-49)^2}{49} = \frac{15^2}{49} = \frac{225}{49} = 4.6$$

The sum of boxes a, b, c and d is:

$$\frac{225}{49} + \frac{225}{31} + \frac{225}{49} + \frac{225}{31} = 23.7$$

The number of degrees of freedom are calculated as:

$$n = \text{(number of rows} - 1) \times \text{(number of columns} - 1) \text{ or } (2-1) \times (2-1) = 1$$

If the calculated value of chi-squared is greater than the critical value from the χ^2 table, then the null hypothesis can be rejected. Obviously 23.7 is greater than 3.84; this is written as $\chi^2 > 3.84$.

So at the $p = 0.05$ level, the null hypothesis can be rejected and we can say that there is an association between the habitats of the two species. There are significantly more quadrats where both occur together, or where neither occurs, than could be accounted for by mere chance. So here the chi-squared test can be used as a **test of association** between the species.

Such a test where we arrange the results in a table with 2 rows and 2 columns is known as a **2 × 2 contingency table**.

Test Yourself

Exercise 11.2.1

1 In a course-work project, it was noticed that there were two types of lichen growing on the stones in an old church graveyard. One was grey in colour and the other yellow. As part of his work, a student observed frequencies of presence:

yellow present, grey present	92
yellow present, grey absent	41
yellow absent, grey present	58
yellow absent, grey absent	34

(a) Make out a 2 × 2 contingency table.
(b) Calculate the expected values for each box in the table.
(c) Work out χ^2 and compare the value that you obtain for one degree of freedom at the appropriate level.
(d) Does this result suggest an association?

2 The World Health Organisation (WHO) investigated an enteric disease in Papua New Guinea to see if it was connected with eating meat there. In the sample that they tested: they found that of the meat-eaters 50 suffered from enteritis necroticans, and 16 did not; of those who didn't eat meat, the sufferers numbered 11 and the non-sufferers, 41.

The null hypothesis is that the two variables are not related. Make a 2 × 2 table and calculate the expected values and then χ^2. Could they assume from their results that there is an association between meat-eating and suffering from the disease?

Causal relationships or not?

Don't forget that these tests of association, although they indicate a link, do not prove that 'x causes y'. The general public sometimes reads much more into statistics than they should do. One survey showed that there is an association between presence of freckles and skin cancer. You can imagine what some of the newspapers made of that piece of information! People became alarmed – yet less than 0.03% of people with freckles get skin cancer (for unfreckled people, it is just above 0.01%). Remember that even though a relationship can sometimes be indicated, it is not always a **causal relationship**. Both variables could be caused by some other variable.

Being able to deal with the 2 × 2 contingency table is much more likely to be of value to you in an investigative project than in an examination question. The same is true for the next technique that demonstrates how to deal with data collected in an ecological study.

χ^2 and ecology

If you wanted to compare the amount of growth of a species in two contrasting habitats (say shaded and open woodland, or exposed and sheltered rocky shores) you could not use raw measurement details (such as areas of leaves or lengths of seaweed fronds). That would be using continuous data and χ^2 can only deal with categories. However, if you scored your observations in terms of frequencies in clearly defined categories – you could use a χ^2 test. The two columns (Table 3) would detail Site 1 and Site 2 and the rows would give the numbers found in each of the size categories in this example.

Example

David was working at a site in South Devon where there were numerous caddis fly larvae in the streams. His Site A was a slow moving shallow stream (0.25 m s^{-1}) and Site B was a much faster stream (0.40 m s^{-1}). The species of caddis that he was studying lays eggs in the water and the larvae build a protective case around the body using particles of sand and small rock particles. He collected 30 specimens at each site and weighed each one.

Fig. 2 Caddis fly larva, 1–3 cm

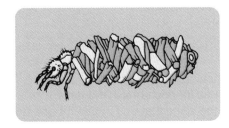

Table 3 *Frequencies of mass categories of larvae*

Mass/g	Site A	Site B
< 0.01	7	1
0.01–0.02	6	2
0.01–0.03	7	5
0.03–0.04	9	11
> 0.04	1	11

To carry out a suitable chi-squared test on these data (which are now in categories), we can assume that the numbers in the categories at Site A are the expected (E) values and the numbers at Site B are the observed (O) values. So substituting in the equation:

$$\chi^2 = \Sigma \frac{(O - E)^2}{E}$$

we have:

$$\frac{(1 - 7)^2}{7} + \frac{(2 - 6)^2}{6} + \frac{(5 - 7)^2}{7} + \frac{(11 - 9)^2}{9} + \frac{(11 - 1)^2}{1}$$

So $\chi^2 = 108.83$ and number of degrees of freedom = $(5 - 1) \times (2 - 1) = 4$

The calculated value is much greater than the critical value for $p = 0.05$, so the null hypothesis (that there is no difference) can be safely rejected. There *is* a difference in the masses of (caddis fly larvae + cases) at the two sites. Of course, we haven't proved that this is due to speed of flow of the streams – it could be because of different availability of food, or oxygen levels, or the materials available for building the cases, or other issues not immediately apparent. All we can say is that there is a difference and it may be worth further investigation.

HINT *Of course, some of the expected values here are less than 5 (see p. 137). In fact, for both sites the values are observed. However, in order to make a comparison we have to make a start by labelling one the expected and the other the observed.*

Exam Questions

Exam type questions to test understanding of Chapter 9

1 In the fruit fly, *Drosophila melanogaster*, the allele for grey body colour, G, is dominant to that for ebony body colour, g. The allele for normal wing, N, is dominant to that for curled wing, n. A student crossed a grey-bodied, normal-winged fly with an ebony-bodied, curled-wing fly. The numbers of the offspring phenotypes were as follows:

Grey body and normal wings	33
Grey body and curled wings	23
Ebony body and curled wings	28
Ebony body and normal wings	16

(a) Show how this cross gave a 1 : 1 : 1 : 1 ratio.
(b) Complete a table showing O (numbers actually observed), E (expected numbers), the difference ($O - E$) and the difference squared $(O - E)^2$.
(c) Calculate the value of χ^2.
(d) Use the χ^2 table to decide whether the observed numbers of offspring are significantly different from those expected. Explain how you reached your answer.

(Adapted from, AQA, 079/1, 1997)

2 Body colour in *Drosophila melanogaster* is influenced by a gene with two alleles: wild type (dominant) and ebony (recessive). A cross was made between two flies, both heterozygous for this gene. The offspring consisted of 40 wild type and 8 ebony-bodied flies. The expected ratio of these offspring is 3 wild type : 1 ebony bodied.

(a) Use this ratio to calculate the expected numbers of each phenotype.

(b) For this cross, the null hypothesis states that there is no significant difference between the results obtained and those predicted by the 3 : 1 ratio. This hypothesis can be tested by calculating the value of chi-squared (χ^2) using the formula below:

$$\chi^2 = \Sigma \frac{(O-E)^2}{E}$$

Tabulate your results in a form suitable for the calculation of (χ^2).

(c) Calculate the value of χ^2, showing stages of working.

(d) Use the table of critical values of χ^2. Does your calculated value of χ^2 enable you to accept or reject the null hypothesis? Explain your answer.

(Adapted from Edexcel, WTA1, June 1997)

3 In a budgerigar cross, the offspring had four different feather colour types, expected to be in the ratio 1 : 1 : 1 : 1. The actual observation was

Dark-green	6
Light-green	10
Cobalt	15
Sky-blue	9

The χ^2 test was used to test whether the observed results fitted the expectation of a 1 : 1 : 1 : 1 ratio. Make a table to show the stages of the calculation.

Calculate the value of χ^2 using the formula:

$$\chi^2 = \sum \frac{(O-E)^2}{E}$$

Use the 5% level of probability in the χ^2 table to explain if the results significantly differed from the expected 1 : 1 : 1 : 1 ratio.

(Adapted from AQA, 0607/1, Summer 2001)

Answers to Test Yourself Questions

Exercise 11.1.1, *p. 136*

1 (a) The total number observed is 546, so the expected numbers are 307.13, 102.38, 102.38 and 34.13. The calculated value of χ^2 is 0.66.

(b) There are 4 categories so 3 degrees of freedom. From the χ^2 table, the critical value at $p = 0.05$ is 7.81. So as the calculated value of 0.66 is so much below the critical value, the ratio can be accepted.

Exercise 11.1.2, *p. 137*

1 There are 38 observed values in 2 categories. If there is no difference in the results, we would expect 19 to have healed in each category. The calculated value of χ^2 is 2.6. This is less than the critical value of 3.84, so we are unable to reject the null hypothesis. So there is a significant difference.

Exercise 11.2.1, *p. 139*

1 Expected values: 88.7, 61.3, 44.3 and ... .7 $\chi^2 = 0.90$ and is less than the critical value, so we must accept the null hypothesis and state that there is no association between the two lichens. *(Some students feel that the investigation has failed if they don't prove an association – they shouldn't – they have succeeded in testing the hypothesis.)*

2 Expected values: 34.1, 26.9, 31.9 and 25.1 value of $\chi^2 = 34.8$, i.e. much greater than the critical value so the null hypothesis must be rejected. There is an association between meat eating and this disease in Papua New Guinea.

Chapter 12

t-test, *U*-test and *D*-test

After completing this chapter you should be able to:

- *carry out calculations of the 'Student's t-test'*
- *check the table of critical values of t at the appropriate level of probability*
- *carry out the U-test and check the critical values*
- *calculate a species diversity index (the D-test).*

12.1 Comparing samples – the 'Student's *t*-test'

Fig. 6 (p. 120) showed a distribution curve for each of two sample populations that were not overlapping. They were completely different in the distribution of the variable being plotted. However *Figs. 7* and *8*, pp. 120–1, each display an overlap in the data – so there is some similarity. The two samples are not identical, but how similar are they? Are the differences large enough to be taken seriously (i.e. are they **significant**?)? We need a *test of the significance of differences* in the population means – one which allows us to compare two sets of observations.

There is such a test that depends on comparing small samples (ideally 30 or fewer) that are normally distributed (though, for work at this level, this assumption can usually be made).

Students often decide to compare populations from two different sites (e.g. normal soil/waterlogged soil; stem growth in sun/shade) or organisms being subjected to two different treatments (e.g. germination times at 10 °C/15 °C; algal cultures at high/low nitrate levels). As usual the first stage of the investigation is to state a null hypothesis, such as 'there is no difference between graph A and graph B' (or even as $\mu_A = \mu_B$).

The test most often used is known as the '**Student's *t*-test**'. Before looking at the formidable equation, it is interesting to note how this test got its name. In 1908, a biologist working in the Guinness brewery in Dublin worked out this useful statistical test. In those days, Mr W.S. Gossett was not allowed by his employers to publish anything under his own name. However he did publish it – but anonymously, using the name 'Student'.

There are versions of this test used under different circumstances and with different-sized samples; but it is acceptable to use the version that follows. If a question is set in examinations, you would not be expected to remember the formula. It would be given, as in the sample questions on p. 151. The test statistic *t* involves the means and standard deviations of the two sets of data and can be checked in a printed table of critical value.

Here is the most usual version of the formula:

$$t = \frac{|\bar{x}_A - \bar{x}_B|}{\sqrt{\dfrac{(S_A)^2}{n_A} + \dfrac{(S_B)^2}{n_B}}}$$

where $|\bar{x}_A - \bar{x}_B|$ is the difference in mean values of sample A and sample B. The vertical lines mean that the sign of the difference is not relevant. We just subtract the smaller from the larger number and ignore the sign.

$(S_A)^2$ and $(S_B)^2$ are the squares of the standard deviations of the samples and n_A and n_B are the sample sizes.

Example

Now let us deal with some data obtained by Open University students, who measured the lengths of leaves in 3-day germinated wheat seedlings that had been given different treatments. Batch A were grown from normal seeds and batch B from seeds that had been subjected to γ-radiation; here are their results:

	Normal, batch A	γ-irradiated, batch B
$\bar{x}$ mean leaf length/mm	10.9	2.3
S standard deviation/mm	3.97	1.52
n sample size	15	15

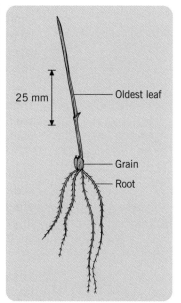

25 mm — Oldest leaf

— Grain
— Root

Fig. 1 A germinating wheat seedling – after 7 days

Substitute these values in the equation:

$$t = \frac{|10.9 - 2.31|}{\sqrt{\left(\dfrac{15.76}{15}\right) + \left(\dfrac{2.31}{15}\right)}}$$

$$= \frac{8.6}{\sqrt{1.19}} = \frac{8.6}{1.09} = 7.89 \quad \text{(to 2 d.p.)}$$

In Table 1 the critical value for t at the 0.05 level and with $(15 - 1) + (15 - 1) = 28$ degrees of freedom is 2.048. So the **probability** of getting a value of t at least as large as 7.89 is less than 0.05 – in fact it is much less than 0.01. So it is extremely unlikely that the difference in these two sets of data could have arisen by chance. We can reject the null hypothesis and describe the difference in the means of A and B as being highly significant.

Table 1 *The t-test for matched and unmatched samples* showing critical values of t at various significance levels. Reject the null hypothesis if your value of t is larger than the tabulated value at the chosen significance levels for the calculated number of degrees of freedom

Degrees of freedom	Significance level					
	20% (0.20)	10% (0.10)	5% (0.05)	2% (0.02)	1% (0.01)	0.1% (0.001)
1	3.078	6.314	12.706	31.821	63.657	636.619
2	1.886	2.920	4.303	6.965	9.925	31.598
3	1.638	2.353	3.182	4.541	5.841	12.941
4	1.533	2.132	2.776	3.747	4.604	8.610
5	1.476	2.015	2.571	3.365	4.032	6.859
6	1.440	1.943	2.447	3.143	3.707	5.959
7	1.415	1.895	2.365	2.998	3.499	5.405
8	1.397	1.860	2.306	2.896	3.355	5.041
9	1.383	1.833	2.262	2.821	3.250	4.781
10	1.372	1.812	2.228	2.764	3.169	4.587
11	1.363	1.796	2.201	2.718	3.106	4.437
12	1.356	1.782	2.179	2.681	3.055	4.318
13	1.350	1.771	2.160	2.650	3.012	4.221
14	1.345	1.761	2.145	2.624	2.977	4.140
15	1.341	1.753	2.131	2.602	2.947	4.073
16	1.337	1.746	2.120	2.583	2.921	4.015
17	1.333	1.740	2.110	2.567	2.898	3.965
18	1.330	1.734	2.101	2.552	2.878	3.922
19	1.328	1.729	2.093	2.539	2.861	3.883
20	1.325	1.725	2.086	2.528	2.845	3.850
21	1.323	1.721	2.080	2.518	2.831	3.819
22	1.321	1.717	2.074	2.508	2.819	3.792
23	1.319	1.714	2.069	2.500	2.807	3.767
24	1.318	1.711	2.064	2.492	2.797	3.745
25	1.316	1.708	2.060	2.485	2.787	3.725
26	1.315	1.706	2.056	2.479	2.779	3.707
27	1.314	1.703	2.052	2.473	2.771	3.690
28	1.313	1.701	2.048	2.467	2.763	3.674
29	1.311	1.699	2.043	2.462	2.756	3.659
30	1.310	1.697	2.042	2.457	2.750	3.646
40	1.303	1.684	2.021	2.423	2.704	3.551
60	1.296	1.671	2.000	2.390	2.660	3.460
120	1.289	1.658	1.980	2.158	2.617	3.373
∝	1.282	1.645	1.960	2.326	2.576	3.291

Table 2 *Probability that chance alone above could produce the difference between observed and expected results*

Degrees of freedom	Significance level				
	0.1	0.05	0.02	0.01	0.001
18	1.734	2.101	2.552	2.878	3.922
28	1.701	2.048	2.467	2.763	3.674
	↑	↑	←↑→		←
	Regarded as *not significant* – any difference could be due to chance.	Difference so great that this could happen by chance only 5 times out of 100.	We can be *confident* that there is a highly significant difference.		We can be *very confident* that there is a highly significant difference.

Test Yourself

Exercise 12.1.1

1 A market gardener was testing the effectiveness of plastic plant pots over clay pots. He used seed from a pure inbred line – so all seeds were the same genotype. He grew 10 plants in plastic pots and 10 plants in clay pots and observed how long it took before each reached a flowering stage suitable for sale. Here are the results.

	A – clay	B – plastic
Number (n)	10	10
Mean of time/days ($\bar{x}$)	95	100
Standard deviation/days (S)	3.2	4.6

State a null hypothesis.

Calculate the value of t and compare it with the values in the table at the appropriate probability level and the correct number of degrees of freedom.

At this stage, you should probably attempt some exam-type questions using the 'Student's t-test' just to reinforce your understanding of the technique.

12.2 The Mann–Whitney *U*-test

The *t*-test is fine if there is a normal distribution; but what if an investigation produces data that do not fit a normal distribution pattern? To compare the two sets (each of which need to have between 6 and 20 values) we could select the **Mann–Whitney *U*-test** (these two statisticians obviously didn't need to remain anonymous!). The principle is that the data from the two samples (either ordinal or interval) are arranged in order of increasing size. Here is a simple example.

Example
A batch of oranges was bought from two shops and each specimen was weighed. Those from shop 1 weighed (in grams) 170, 170, 180, 185, 200, 207 and 220.

The oranges from shop 2 weighed 175, 178, 205, 210, 210, and 210. The first stage in the calculation is to arrange them all in *rank order* (plotted along a single line).

Shop 1/g	170	170			180	185	200		207				220
Order	1	2	3	4	5	6	7	8	9	10	11	12	13
Shop 2/g			175	178				205		210	210	210	

We can note that the 13 oranges are spread between the two sites (shops) and that those from one shop are not all bigger than those from the other. We also note that some of them are the same size, so that in the rank order the mass of 170 g is in positions 1 and 2. So we take the mean of 1 and 2 and give them both an equal rank of 1.5. It's the same with 210 g – the mean of positions 10, 11 and 12 is 11 and the next one, 220 g must be in position 13. So we can re-write the order as:

Shop 1/g	170	170			180	185	200		207				220
Rank order, *R*	1.5	1.5	3	4	5	6	7	8	9	11	11	11	13
Shop 2/g			175	178				205		210	210	210	

Note where the median values are in each set – 185 g in shop 1 and between 205 and 210 g in shop 2). There is obviously some overlap, but there is no need to look for a normal distribution. We can use the distribution free test (also known as a **non-parametric test**), the Mann–Whitney *U*-test. This is the technique.

Method
Add together the rank positions occupied in each set to give ΣR, so for shop 1:

$$\Sigma R_1 = 1.5 + 1.5 + 5 + 6 + 7 + 9 + 13 = 43$$

and for shop 2:

$$\Sigma R_2 = 3 + 4 + 8 + 11 + 11 + 11 = 48$$

KEY FACT The value of the *U*-test statistic is, for site 1:

$$U_1 = (n_1 \times n_2) + \frac{n_2(n_2 + 1)}{2} - \Sigma R_2$$

and for site 2:

$$U_2 = (n_1 \times n_2) + \frac{n_1(n_1 + 1)}{2} - \Sigma R_1$$

n_1 and n_2 are the number of oranges in each set ($n_1 = 7$ and $n_2 = 6$).

Substituting our values into the equation we have:

$$U_1 = (7 \times 6) + \frac{6(6 + 1)}{2} - 48 = 15$$

and

$$U_2 = (7 \times 6) + \frac{7(7 + 1)}{2} - 43 = 27$$

The next stage is to take the smaller U-value (in this case, 15) and to compare it in Table 3 with the critical value (using the values of n_1 and n_2 to find the critical value). Using 6 and 7, we see a critical value of 6.

Table 3 *Critical values for the Mann–Whitney U-test* (at the $p = 0.05$ *level). If the smallest U-value is less than or equal to the critical value, then there is a significant difference between the two sets of data – so reject the null hypothesis (no decision possible)*

	Values of n_2																			
	1	2	3	4	5	6	7	8	9	10	11	12	13	14	15	16	17	18	19	20
1	—	—	—	—	—	—	—	—	—	—	—	—	—	—	—	—	—	—	—	—
2	—	—	—	—	—	—	—	0	0	0	0	1	1	1	1	1	2	2	2	2
3	—	—	—	—	0	1	1	2	2	3	3	4	4	5	5	6	6	7	7	8
4	—	—	—	0	1	2	3	4	4	5	6	7	8	9	10	11	11	12	13	13
5	—	—	0	1	2	3	5	6	7	8	9	11	12	13	14	15	17	18	19	20
6	—	—	1	2	3	5	6	8	10	11	13	14	16	17	19	21	22	24	25	27
7	—	—	1	3	5	6	8	10	12	14	16	18	20	22	24	26	28	30	32	34
8	—	0	2	4	6	8	10	13	15	17	19	22	24	26	29	31	34	36	38	41
9	—	0	2	4	7	10	12	15	17	20	23	26	28	31	34	37	39	42	45	48
10	—	0	3	5	8	11	14	17	20	23	26	29	33	36	39	42	45	48	52	55
11	—	0	3	6	9	13	16	19	23	26	30	33	37	40	44	47	51	55	58	62
12	—	1	4	7	11	14	18	22	26	29	33	37	41	45	49	53	57	61	65	69
13	—	1	4	8	12	16	20	24	28	33	37	41	45	50	54	59	63	67	72	76
14	—	1	5	9	13	17	22	26	31	36	40	45	50	55	59	64	67	74	78	83
15	—	1	5	10	14	19	24	29	34	39	44	49	54	59	64	70	75	80	85	90
16	—	1	6	11	15	21	26	31	37	42	47	53	59	64	70	75	81	86	92	98
17	—	2	6	11	17	22	28	34	39	45	51	57	63	67	75	81	87	93	99	105
18	—	2	7	12	18	24	30	36	42	48	55	61	67	74	80	86	93	99	106	112
19	—	2	7	13	19	25	32	38	45	52	58	65	72	78	85	92	99	106	113	119
20	—	2	8	13	20	27	34	41	48	55	62	69	76	83	90	98	105	112	119	127

(left axis label: Values of n_1)

The rule is that if the smaller U value (in the example $U_1 = 15$) is equal to or less than the table critical value, we can reject the null hypothesis that there is no difference; accepting that there is a significant difference between the two medians. As our value of U is more than the critical value, we accept the null hypothesis that there is no significant difference in the values based on the medians of the oranges.

Test Yourself

1 Two students were comparing collections of larvae of a caddis fly that builds its case from tiny snail shells. They obtained these values by using a *kick sampling technique* (a time of 1 min collecting for each sample). In the first stream they collected 8 samples: 1, 0, 5, 10, 4, 12, 7, 14. In the second stream they had time to collect only 7 samples: 15, 13, 8, 13, 20, 20, 19.

Calculate the values of *U* and use the table to decide if (at $p > 0.05$) there is a significant difference in the two streams.

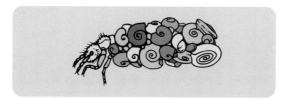

Fig. 2 *Caddis fly larva, 1–3 cm*

12.3 Simpson's diversity index, *D*

You already know that the abundance of individual species in an investigation area could be measured on a scale such as the ACFOR scale (see p. 73). One other technique which may be of value to you in ecological studies is to describe the *richness* of different habitats. It is obvious that, with modern agriculture, you would expect to find just one plant species dominant in a field. So that potatoes, or wheat or carrots would be found with, perhaps, a small number of weeds of a few species that had escaped the pesticide spray. Contrast this with an old-fashioned meadow that had received no attention at all other than from some grazing cattle. There, you would find a huge number of many different wild species (as well as the main agricultural crop – grass forage). This type of meadow is described as being *species rich*.

Much of the work of professional ecologists today is to give opinions in terms of conservation value on the species richness of an area and the term **diversity** describes this.

There are a number of different methods that can be used and some very complicated formulae that really need a good spreadsheet to aid calculation. However, we most often use, at this level, a straightforward one giving an *index* (a single number). This brings together the number of species (or it could be any other taxonomic grouping, such as plant families or invertebrate genera) with the number of individuals found from each taxonomic group. It is easier to explain this by following an example of the calculation of **Simpson's diversity index**.

Example

Let us assume that on one side of a wall there is a conservation site, area 1, and on the other side a managed grouse moor, area 2. On each side of the wall, 100 plants were pinpointed in a random method and identified. It doesn't matter if you can't name all of the species, but you must be able to note the different ones present. In this example, the plants are numbered 1–8.

Species	Area 1		Area 2	
	n_1, number of plants	$n_1(n_1 - 1)$	n_2, number of plants	$n_2(n_2 - 1)$
1	35	$35 \times 34 = 1190$	74	$74 \times 73 = 5402$
2	14	$14 \times 13 = 182$	20	$20 \times 19 = 380$
3	13	$13 \times 12 = 156$	3	$3 \times 2 = 6$
4	12	$12 \times 11 = 132$	2	$2 \times 1 = 2$
5	8	$8 \times 7 = 56$	1	$1 \times 0 = 0$
6	7	$7 \times 6 = 42$	0	
7	6	$6 \times 5 = 30$	0	
8	5	$5 \times 4 = 20$	0	
	$\sum n_1 = N_1 = 100$	$\sum n_1(n_1 - 1) = 1808$	$\sum n_2 = N_2 = 100$	$\sum n_2(n_2 - 1) = 5790$

KEY FACT

Simpson's diversity index, $D = \dfrac{N(N - 1)}{\sum n(n - 1)}$

Where N = total number of organisms of all species and
n = total number of organisms of a particular species.

For the first site:

$$D_1 = \frac{(100 \times 99)}{1808} = 5.5$$

and for the second site:

$$D_2 = \frac{(100 \times 99)}{5790} = 1.7$$

These values are not absolute, they are just used for comparison; a higher value of D suggests higher diversity. In this example, the conservation area has a higher species diversity index than the managed grouse moor. A clean freshwater stream could be expected to have a higher diversity index of invertebrate fauna than a polluted one. Managed conifer plantations would have a lower diversity index of ground flora species than natural oak woodland. In fact, the diversity index is often used as a measure of human interference in a habitat.

Exam Questions

Exam type questions for understanding of Chapter 12

1 A freshwater arthropod *Gammarus pulex* may be infected by a parasite. 20 infected and 20 non-infected *G. pulex* were placed in an aquarium tank and observed over a period of 5 min to see how long they spent at the surface.

	Mean time near surface/s	Standard deviation/(S)
Infected *G. pulex*	46.8	7.8
Non-infected *G. pulex*	21.5	3.7

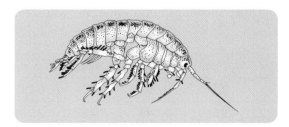

Fig. 3 *The arthropod **Gammarus pulex***

(a) Explain why it is appropriate to use a *t*-test to analyse these data.
(b) The formula used for the *t*-test is:

$$t = \frac{|\bar{x}_A - \bar{x}_B|}{\sqrt{\dfrac{(S_A)^2}{n_A} + \dfrac{(S_B)^2}{n_B}}}$$

x_A is the mean time near the surface for the infected population
x_B is the mean time near the surface for the non-infected population
S_A and S_B are the standard deviations and
n_A and n_B are each 20, the number of organisms in each population.

Use the values of S from the table to calculate $(S_A)^2$ and $(S_B)^2$.
Use the other values to calculate the value of t. Show your working.
This extract from a statistical table shows the values of t for 38 degrees of freedom.
Using your value for t and your null hypothesis make a statement about the behaviour of the infected and non-infected *G. pulex.*

(Adapted from Edexcel, 6045/01, January 2001)

Significance levels						
d.o.f.	0.1	0.05	0.02	0.01	0.002	0.001
38	1.686	2.024	2.429	2.712	3.319	3.566

2 Pairs of the woodland bird *Parus major* were observed at two sites, A and B. Each pair produced two broods of young each year; the number of days between the two broods was recorded.

Site	Number of samples – pairs (*n*)	Mean number of days between broods ($\bar{x}$)	Standard deviation (S)
A	18	39.8	1.34
B	11	32.5	3.18

(a) A suitable null hypothesis would be that there is no difference between the mean number of days between broods for the two sites. A *t*-test can be made using the formula below.

$$t = \frac{|\bar{x}_A - \bar{x}_B|}{\sqrt{\dfrac{(S_A)^2}{n_A} + \dfrac{(S_B)^2}{n_B}}}$$

Calculate the value of *t* and show your working.

(b) An extract from a table showing critical values of t with 27 degrees of freedom is given below.

d.o.f	Significance levels			
	0.1	0.05	0.02	0.01
27	1.699	2.045	2.462	2.756

Using this information and your calculated value of t, what conclusions can you draw about the results of this investigation?

(Adapted from Edexcel, HB6, January 2001)

3 An investigation involved a study of the diversity of lichens on tree trunks. The results for one tree trunk, 8 km from the city centre are shown.

Species of lichen	Number of individuals
A	12
B	3
C	3
D	2

The index of diversity can be calculated from the formula:

$$\text{Index of diversity} = \frac{N(N-1)}{\sum n(n-1)}$$

where N = total number of organisms of all species and
n = total number of organisms of a particular species

Use this formula to calculate the index of diversity for the lichens on this tree trunk. Describe and explain how you would expect the index of diversity of lichens on tree trunks to change as you moved towards the city centre.

Exercise 12.1.1, *p. 146*
The null hypothesis is that there is no difference in the mean time taken to reach sale condition whether plastic or clay pots are used.
The value $t = 2.82$ and, from the table, the critical value for the 0.05 level and 18 degrees of freedom is 2.101. So, as the calculated value is greater, the null hypothesis can be rejected and the gardener may consider it is better to use clay pots (though they may be more expensive!).

Exercise 12.2.1, *p. 149*
$U_1 = 0$; $U_2 = 41.5$. The critical value from the table is 10, so the U_1 value of 0 is less and we can assume a significant difference at the $p = 0.05$ level.

Appendix

Critical values of χ^2 at various significance levels. Reject the null hypothesis if your value of χ^2 is bigger than the tabulated value at the chosen significance level, for the calculated number of degrees of freedom.

degrees of freedom	Significance level				
	$p = 0.2$ 20%	0.1 10%	0.05 5%	0.02 2%	0.01 1%
1	1.642	2.706	3.841	5.412	6.635
2	3.219	4.605	5.991	7.824	9.210
3	4.642	6.251	7.815	9.837	11.341
4	5.989	7.779	9.488	11.668	13.277
5	7.289	9.236	1.070	13.388	15.086
6	8.558	10.645	12.592	15.033	16.812
7	9.803	12.017	14.067	16.622	18.475
8	11.030	13.362	15.507	18.168	20.090
9	12.242	14.684	16.919	19.679	21.666
10	13.442	15.987	18.307	21.161	23.209
11	14.631	17.275	19.675	22.618	24.725
12	15.812	18.549	21.026	24.054	26.217
13	16.985	19.812	22.362	25.472	27.688
14	18.151	21.064	23.685	26.873	29.141
15	19.311	22.307	24.996	28.259	30.578
16	20.465	23.542	26.296	29.633	32.000
17	21.615	24.769	27.587	30.995	33.409
18	22.760	25.989	28.869	32.346	34.805
19	23.900	27.204	30.144	33.687	36.191
20	25.038	28.412	31.410	35.020	37.566
21	26.171	29.615	32.671	36.343	38.932
22	27.301	30.813	33.924	37.659	40.289
23	28.429	32.007	35.172	38.968	41.638
24	29.553	33.196	36.415	40.270	42.980
25	30.675	34.382	37.652	41.566	44.314
26	31.795	35.563	38.885	42.856	45.642
27	32.912	36.741	40.113	44.140	46.963
28	34.027	37.916	41.337	45.419	48.278
29	35.139	39.087	42.557	46.693	49.588
30	36.250	40.256	43.773	47.962	50.892

Answers to exam-type questions

Chapter 1

1 (a) Adults: 1.2% by number; 43.7% by energy
      value
      Nestlings: 1.3% by number; 23.9% by energy
      value
   (b) Caddis larvae: 68.6%

Chapter 2

1 The measurement is between 11 and 12 mm.

   $11 \div 12\,000 = 0.92\ \mu m$ **or** $12 \div 12\,000 = 1.0\ \mu m$

2 $\dfrac{418.0 - 26.8}{418} \times 100 = \dfrac{391.2}{418} \times 100$

   or $100 - \dfrac{26.8}{418} \times 100$

   Answer is 93.58 to 93.6% (*not 94 – one mark for working and one mark for accurate answer*)

3 In this type of answer you must refer to the numbers and units in the question. For example, you might refer to:
   'The uptake remains constant at $4\ cm^3 h^{-1}$ for the first 6 h. After this time there is a distinct/sharp increase over $4\frac{1}{2}$ h, reaching a maximum of $62\ cm^3 h^{-1}$. This is then followed by a steep decline, with periods of no change at intervals around early evening (15.00 to 17.00 and 18.30 to 21.00). (*Note also an interesting feature of this graph – there are two labelled y-axes, but both refer to the same x-axis.*)

Chapter 3

1 104 (Did you remember to find the size of the whole second sample?)
2 270
3 (a) 45 (b) 16 (c) 2 (d) 15 (e) 121 (f) 4
4 Total population = 233 – so black were 36% of the total. Therefore $q^2 = 0.36$ and $q = \sqrt{0.36} = 0.6$;
   $p = 1 - 0.6 = 0.4$; $2pq = 2 \times 0.4 \times 0.6 = 0.48$;
   $0.48(149 + 84) = 112$

Chapter 4

1 (a) 385 cows (*Don't forget to show your workings, because such questions often give one mark for correct workings and one mark for correct answer.*)
   (b) AI is more efficient than the natural method (385 rather than 1!); makes more use of good breeding stock; is economical; has a higher success rate.
2 158 cells; 165 µm

HINT *Start by converting the lengths to the SI unit – the metre.*

Remember that: $1\ m = 10^3\ mm = 10^6\ \mu m$

So $19\ mm = 19 \times 10^{-3} m$ *and* $120\ \mu m = 120 \times 10^{-6}\ m$)

Chapter 5

1 For worm: $\dfrac{\text{water lost}}{2.5} = 2.61$
   so water lost = $6.53\ cm^3\,g^{-1}\,min^{-1}$
   For insect: $\dfrac{\text{water lost}}{2.5} = 0.11$
   so water lost = $0.28\ cm^3\,g^{-1}\,min^{-1}$

Chapter 6

1 The graph should show: axes right way named and correctly labelled with units; suitable scale, at least half the graph paper used; points accurately plotted; points joined using ruled lines, no extrapolation to O (the origin); key to curves.

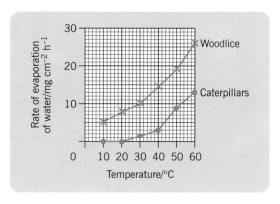

Evaporation from the surface of two anthropods

2 The graph should show: axes correct and labelled with units; scale arithmetic, using at least half the graph paper; points plotted accurately; curve drawn by joining successive points by ruled lines; key for curves.

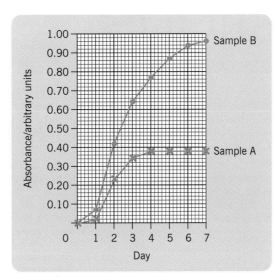

Growth of **S. capricornutum**

3 (a) 4.0 ms^{-1} (b) Group E
   (c) No, anaerobically, because at that speed both 100 m and 440 m athletes are anaerobic; so it is likely that 200 m athletes will also be anaerobic.

Chapter 7

1 (a) (i) J 2.5–3.0%;   K 6%;   (2)
      (ii) *Marks are given for*:
      CO_2 decreases/or uptake increases as light increases; gradient steepest up to 5%; little change after 15%/(nett) CO_2 produced/or released up to 2.5/3%; below 3.0%/converse; correct ref. to relative rate of photosynthesis and respiration; correct ref. to compensation point;   (3)
   (b) (i) *Any two of these points allowed.*
      gradient steeper/levels off (at lower light intensity)/reaches a maximum; maximum rate of uptake is lower/or 33 units difference/higher uptake in K at high light intensities; change not directly proportional to light intensity; J has a higher uptake below 19.5%; J reaches its compensation point at lower light intensities than K; *Allow converse if K stated.*   (2)
2 32 cells after 13 h; 32/0.004 mm^3 = 8000 in 1 mm^3; 800 000 or 8 × 10^6 in 1 cm^3

Chapter 8

1 r_s = 0.277
2 r_s = 0.089

Chapter 9

1 Even from the table it looks like a normal distribution – but you could do a quick sketch to check and to demonstrate that you understand.
2 (b) The class that has the greatest number of individuals measured in the sample (the mode) is 100–9 mm. (*Don't forget to state the units in any answers.*)
   (c) As there were 164 specimens in the sample, the median would be the middle one – but in this question, with an even number of specimens, it would be the mean of the 82nd and the 83rd (which would each measure somewhere between 110 mm and 119 mm. So the median value would be somewhere between 110 mm and 119 mm.

Chapter 10

1 *a short answer question* (a) 70–130 (*1 mark*)   (b) 2.5% (*1 mark*)
2 You should explain the terms mean and standard deviation (2 marks). Then give an indication of the percentages that you would expect to find within the limits of one and two standard deviations above and below the mean (*3 marks*).
3 (a) 161–187 cm   (b) 154.5–193.5 cm
4 Mean = 35.9 mm and standard deviation = 4.32 mm
5 Mean of A = 18.53 cm^2 and standard deviation of A = 2.87 cm^2
   Mean of B = 15.7 cm^2 and standard deviation of B = 2.7 cm^2.

Chapter 11

1 (You need to have done the genetics to be able to do part (a).) (c) and (d) χ^2 = 6.32, which is less than the critical value (3 degrees of freedom, p = 0.05) therefore there is no significant difference between observed and expected values.
2 (a) The expected values are 36 and 12. (c) and (d) The value of χ^2 is 1.77 (much less than the critical value) so the null hypothesis can be accepted.
3 The estimated values are 10, 10, 10 and 10. The value of χ^2 = 4.2 (3 degrees of freedom) therefore there is no significant difference.

Chapter 12

1 $(S_A)^2$ = 60.84; $(S_B)^2$ = 13.69; t = 12.52
   This value (at p = 0.05 level) is much higher than the table value of 2.024. So it is extremely unlikely that the difference is due to chance alone. So the null hypothesis is rejected. (*Don't forget that d.o.f. is (20 – 1)+(20 – 1) = 38).*
2 t = 7.23. This is greater than the critical value – so the null hypothesis is rejected.
3 D = 2.60; the value of D would decrease because of pollution.

Further practice questions

1 Leaves which are adapted to low light intensities are known as shade leaves, while those which function more efficiently in high light intensities are known as sun leaves.

An investigation was carried out to determine whether there was a significant difference in the surface area of shade and sun leaves of dog's mercury (*Mercurialis perennis*), a plant which grows in woodland and shady places.

Seventeen leaves of dog's mercury were collected from plants growing in deep shade (site A) and seventeen leaves from plants growing in a clearing open to sunlight (site B). The surface area of each leaf was measured. The results are shown in the table below.

A *t* test was carried out to determine whether the difference in mean surface areas was significant at the 5% level.

Surface area of leaves/cm²	
Deep shade (Site A)	**Open clearing (Site B)**
21	15
14	17
16	18
18	17
19	17
21	19
19	13
22	14
18	21
16	13
13	16
22	13
21	16
23	12
19	14
18	12
15	20
Mean $\bar{x}_A = 18.53$	Mean $\bar{x}_B =$
Standard deviation (A) $S_A = 2.87$	Standard deviation (B) $S_B = 2.70$

The formula used for the *t* test was

$$t = \frac{(\bar{x}_A - \bar{x}_B)\sqrt{(n-1)}}{s}$$

where $\bar{x}_A$ is the mean surface area from site A
      $\bar{x}_B$ is the mean surface area from site B
      n is 17, the number of leaves from each site
      s is found by the formula
$$s^2 = s_A^2 + s_B^2$$

(a) (i) The mean surface area ($\bar{x}_A$) of the leaves from site A is given in the table. Calculate the mean surface area ($\bar{x}_B$) of the leaves from site B. Write your answer in the space in the table.

     (ii) Calculate the value of s^2 and of s. Show your working.

    (iii) Use your values from (a) (i) and (a) (ii) to calculate the value of t. Show your working.

(b) A statistical table showed that the significance at the 5% level with 32 degrees of freedom required a *t* value of at least 2.04. What does this indicate about the difference in mean surface areas between leaves from site A and leaves from site B?

(c) Suggest *one* reason for the differences in surface areas between leaves from the two sites. Explain your answer.

(Edexcel, 6046 (B6), June 1997)

2 A long term investigation was carried out to test the hypothesis that populations of mosquitoes which are regularly sprayed with insecticide will become resistant to it.

A large sample of mosquitoes was randomly collected from a village which had been regularly sprayed with insecticide. 250 of these mosquitoes were placed in a sealed container and a measured dose of insecticide was added. After one hour, the total number of mosquitoes killed by this treatment was counted. The experiment was then repeated four times using fresh batches of mosquitoes and increasing the dose of insecticide each time.

A second sample of mosquitoes of the same species was then collected from a village which had never been sprayed with insecticide and tested in the same way.

An extract from the records of this investigation is shown below.

250 mosquitoes added		
mg insecticide	Sprayed village	Unsprayed village
0.15	45	188
0.25	70	230
0.35	95	245
0.45	102	235
0.55	132	238
	numbers killed	

(a) Calculate the percentage of mosquitoes killed by the different insecticide treatments for each village. Then organise the data into a suitable table so that the effect of increasing insecticide concentration for each village can be compared.
(b) Use the data in your table to present this information in a suitable graphical form.
(c) What conclusions can you draw from the results of this investigation?

(Edexcel, 6048/03, January 2001)

3 Herds of thousands of wildebeest graze the high productivity grasslands of the Serengeti in Tanzania. Some areas were fenced off to prevent grazing and researchers measured the biomass from the time that the migrant herds arrived.
(a) Define the terms **biomass** and **productivity**.
(b) (i) using the data in the table calculate the mean rate of change of fresh biomass per day, for both grazed and ungrazed grassland, between days 1 and 32. Show your working.

Table showing the fresh biomass over a 32 day period on grazed and ungrazed Serengeti grassland. Day 1 is the first day after the wildebeest moved on.

Day	Fresh biomass/g m^{-2}	
	Grazed	Ungrazed
1	50	430
8	55	420
16	100	380
24	120	350
32	200	300

(ii) Suggest why the fresh biomass in the ungrazed area decreased.

(Adapted from Edexcel, 6103/03, June 2001)

4 An experiment to determine the rate of carbohydrate formation and use in a cabbage plant at various temperatures was carried out. The experiment was carried out at a high light intensity. The results are given below:

Temperature/°C	Carbohydrate production (Photosynthesis)/ mg h^{-1}	Carbohydrate use (Respiration)/ mg h^{-1}
0	0	2
5	7	2
10	30	3
15	44	4
20	72	5
25	74	8
30	48	10
40	12	18
50	0	30
60	0	18

(a) Plot the data and clearly label each graph.

Use the graph, together with your own knowledge, to answer questions (b) to (f).

(b) At what temperature do you think most oxygen will be produced?
(c) At what temperature is the rate of carbohydrate production equal to the rate of carbohydrate use?
(d) At what temperature will most carbon dioxide be produced?
(e) At what temperature will the plant grow best?
(f) Estimate carbohydrate use at 65 °C. Why does carbohydrate use decrease at higher temperatures?

5 The distribution of two species of buttercups, *Ranunculus repens* and *Ranunculus bulbosus*, show a marked correlation with drainage. It is suggested that seedling establishment is the stage in the life cycle that is susceptible to the drainage conditions.

In an investigation of the germination and establishment of seedlings of the two species, seeds were sown in pots, some with free drainage and some half-waterlogged (achieved by immersion of the bottom half of the pot in water). In the freely draining pots, 230 of the *R. repens* seedlings and 330 of the *R. bulbosus* seedlings had established a month after sowing. In the half-waterlogged pots 270 of the *R. repens* seedlings and 180 of the *R. bulbosus* seedlings had established in the same time.
(a) Arrange the data for analysis in a 2 × 2 contingency table.
(b) The results were analysed using the 2 × 2 χ^2 test.
   (i) State the appropriate null hypothesis for this test.
   (ii) The χ^2 value was calculated as 35.762. Determine the probability level (p) for this statistic.

(iii) On the basis of the probability level (p) determined in (ii) above, state your decision about the null hypothesis.
(c) What can you conclude regarding the suggestion that seedling establishment is the stage in the life cycle that is susceptible to the drainage conditions.

(CCEA, BM2F0, February 2000)

6 The effect of applying a nitrogen fertiliser to a wheat crop is shown in the graph.

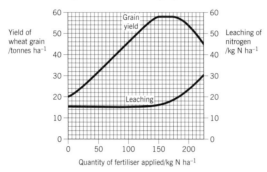

Explain the shape of the curve for
(i) grain yield when fertiliser application increases from 0 to 100 kg per hectare,
(ii) grain yield when fertiliser application increases from 125 to 175 kg per hectare,
(iii) grain yield when fertiliser application increases from 175 to 225 kg per hectare,
(iv) leaching of nitrogen when fertiliser application increases from 150 to 225 kg per hectare.

(AQA, BY07, 1998)

7 The distribution of plants along a section of heathland was investigated using a transect. A table of results was drawn up.

	Percentage cover at metre intervals														
	1	2	3	4	5	6	7	8	9	10	11	12	13	14	15
Sundew													5	10	5
Bracken	30	40	35	30	25	15		5							
Rush						5	5	10	15	40	25				
Bogmoss				10	5				10	15	40	75	60	70
Stonecrop	5	15	5	10	5										

(a) Describe how the data for percentage cover of bracken could have been obtained.
(b) Construct kite diagrams to show the distribution of the five species.

(adapted from AQA, 5523)

8 Moose are large herbivorous animals.

In a study of one population of moose, 72 animals were trapped and marked with ear tags. They were then released. One month later, fieldworkers examined 120 moose and found that 14 of these had ear tags.

Use these figures to calculate the size of the moose population. Show your working.

(From AQA, AS-A/MB10/2, June 1999)

9 An agricultural company is proposing to set up a farm to start breeding bison. They have searched the web for some nutritional details of bison meat and have asked you to devise some graphs to promote the product and which could be used to advertise to the public. Use the data in the table below to draw the most suitable graphs.

	Protein mg per 100g	Fat mg per 100g	Cholesterol mg per 100g
Fish (cod)	20	7	57
Chicken	29	7	90
Beef	30	30	85
Pork	27	16	93
Bison	34	3	40

10 The diagram below shows the quantity of energy flowing through a food chain in a terrestrial ecosystem. The figures given are kJ m^{-2} yr^{-1}.

(a) Calculate the percentage of the incident energy which becomes available as the net primary production (NPP) of green plants. Show your working.
(b) Give *two* reasons why not all the energy of the incident sunlight is incorporated into biomass of green plants.

(Edexcel, 6042(B2), June 1997)

11 Deserts are very inhospitable environments. Water is always in short supply and temperatures during the day may be very high although those at night may be low.
(a) The table shows how the diversity of small mammals varies with mean annual rainfall.

Index of diversity	Mean annual rainfall/cm
1.0	7
1.3	9
1.7	9
1.8	14
1.8	20
2.3	21
2.5	22
2.7	27
3.2	27

(i) Plot these data as a scatter diagram.
(ii) Draw the line of best fit through the points. Use this line to describe the

relationship between species diversity and mean annual rainfall.

(iii) Explain the relationship described in your answer to (a)(ii).

(AQA, 0607/2, June 1999)

12 (a) Table 1 shows the percentage of different bases in DNA from different organisms.

Table 1

Source of DNA	Adenine %	Guanine %	Thymine %	Cytosine %
Human	30	20	30	20
Rat	28	22	28	22
Yeast	31	19	31	19
Turtle	28	22	28	22
E.coli	24			
Salmon	29	21	29	21
Sea urchin	33	17	33	17

(i) What information about the ratios of the different bases in DNA can you work out from the table?

(ii) Give the results that you would expect for DNA from the *E.coli* bacterium. Explain how you arrived at your answer.

(iii) Turtles have the same percentages of the four different bases as rats. Explain why they can still be very different animals.

(b) Table 2 shows the percentage of different bases in the DNA from a virus.

Table 2

Adenine %	Guanine %	Thymine %	Cytosine %
25	24	33	18

(i) Describe how the ratios of the different bases in this virus differ from those in Table 1.

(ii) The structure of the DNA in this virus is not the same as DNA in other organisms. Suggest what this difference in DNA structure might be.

(AQA, BYA2, January 2001)

13 The table shows the number of reported cases of food poisoning caused by *Salmonella enteritidis* and *Salmonella typhimurium* each year in England and Wales during the period from 1982 to 1992.

Year	Number of reported cases of food poisoning by *Salmonella enteritidis*	Number of reported cases of food poisoning by *Salmonella typhimurium*
1982	1 000	5 500
1983	2 050	6 100
1984	2 200	7 150
1985	3 150	5 000
1986	4 000	6 150
1987	6 950	7 000
1988	12 200	6 050
1989	11 850	6 200
1990	16 400	5 000
1991	14 600	4 950
1992	21 300	5 100

(Adapted from Phillips and George, Biologist Volume 41, Number 2 (April 1994))

(a) In 1992, a total of 63 450 cases of food poisoning was reported in England and Wales. Calculate the percentage of this total that was due to infection by the two species *S. enteritidis* and *S. typhimurium*. Show your working.

(b) Compare the trends in the number of cases of food poisoning caused by *S. enteritidis* with those caused by *S. typhimurium* during the period from 1982 to 1992.

(From Edexcel, 6044/01, January 2001)

Index

A

abbreviations *see* symbols
ACFOR scale 73
algebra 36–43
alleles 41–2, 56
alphabet, Greek 16, 36, 134
arbitrary units 80, 82
arithmetic mean 114
association, test of 139, 140
averages 114–17
axes, graphs 68

B

bar charts 70, 74
base units 16, 28–9
base value (indices) 44
bell-shaped curves 119
best fit lines 78, 106, 107, 110
bimodal distributions 120
biological measurement 17–19

C

calculations from graphs 86–8
calculators 11–12, 49, 109–10, 130
calibration 23–5, 96
capital letters 29–30
categorical data 64, 118, 136
causal relationships 103, 140
cell counting 26–7
Celsius 29, 30
centimetres 17
central tendency 114, 116
change, percentage 8–9
changing the subject of equations 37–9
charts 63, 64, 70–4, 79–80, 84
chi-squared test 134–7, 139, 153
chromatography ratios 60–1
co-ordinates 92
colorimeters 82
'commenting on' 88
compound units *see* derived units
confidence limits 131
constants 38
contingency tables 138–9
continuous data 118, 136
correlation
    coefficient 105, 110–13
    definition 112
    negative relationship 102, 105

nil correlation 103, 105
    positive relationship 101, 102, 103, 105
    Spearman's rank correlation test 110–12, 113
counting cells 26–7
critical values 112, 148, 153
curves 68, 78, 96
    see also distributions
    bell-shaped 119
    exponential 90

D

D-test 149–50
data
    correlation 101–13
    display 62–84
    graph interpretation 85–6
    types 64
decimal notation 2, 3, 5, 9
degrees of freedom 135, 137, 139, 145
density, units 31
dependent variables 76, 103
derived units 31–2
dihybrid ratios 54–6
dilution ratios 60
discontinuous variables 70
discrete data 64, 118
displaying data 62–84
distribution free tests 146–9
distributions
    bimodal 120
    confidence 131
    negatively skewed 119, 120
    normal 123–30
    positively skewed 119, 120
    standard deviation 123–30
diversity index (*D*) 149–50
dividing fractions 4
Domin scale 65
double-blind trials 137
drawing graphs 66–84
drug trials 137

E

ecology 40–1, 110–12, 140–1
Egyptian notation 2–3
electron micrographs 19, 20, 21
enzymes 95
equations 36–41

error
    bars 96, 97, 131
    levels 96
    standard 131
estimates, without calculator 10
expected value 134
exponent, indices 44
exponential curves 90
extrapolation 93–5

F

field work 40–1
formulae 39
fractions 3–5
freedom, degrees of 135, 137, 139, 145
frequency distributions 118, 119, 120
frequency units 33

G

Gaussian distribution *see* normal distribution
genetics 41–2, 52–6
Gigametres 17
Gossett, W. S. 143
gradients 95, 96, 108
graphs 2, 62
    co-ordinates 92
    curved 96
    data correlation 101–10
    drawing 67–78
    error levels 96
    extrapolation 93–5
    frequency distributions 119–20
    gradients 95, 108
    intercepts 92, 108
    interpolation 93
    interpreting 85–6
    line of best fit 78, 106, 107, 109, 110
    log scales 12, 18–19, 89–91
    standard deviation 96–7
    straight line graphs 95
    types 64
graticules 23–5
Greek language 15–16, 36, 134

H

H_0 *see* null hypothesis
habitat richness 149–50

haematocytometer slides 26
Hardy–Weinberg equilibrium 41–2
hertz 33
hieroglyphs 2–3
histograms 70–3, 124
history 1

I

improper fractions 4
independent variables 67, 76, 103
indices 44–8
inheritance *see* genetics
intercepts 92, 108
International System of Units (SI) 3, 16, 28
interpolation 93, 109

J

joules 29

K

kelvins 29
kick sampling technique 149
kilometres 17
kite diagrams 72–4, 84

L

Latin language 15–16
length, units 17
level of significance 135, 145, 146, 153
Lincoln Index 41
line of best fit 66, 78, 106, 107, 109, 110
line graphs, plotting 75–8
logarithmic scales 12, 18–19, 89–91
lower case 29–30

M

magnification 20, 21, 22, 59
Mann–Whitney *U*-test 146–9
mark-release-capture technique 41
matched samples 145
mathematical symbols 28
mean 114–33
    arithmetic mean 114
    frequency distributions 119
    measure of central tendency 114, 116
    median 115, 119
    mode 116, 119, 120
measure of central tendency 114
measurement, biological objects 17–19
median 115, 119
Megametres 16, 17
Mendel, Gregor 1, 52, 53, 54, 55, 134
meter/metre 17
metres 17
micrometer scales 23–5
micrometres 17
microscopes 22

calibration 23–5
cell counting 26–7
graticules/micrometers 23–5
resolving power 19–20
millimetres 17
mixed numbers 4
mode 116, 119, 120
moles 29
monohybrid inheritance 54
multiplying fractions 4

N

nanometres 17
nearest whole number 9–10
negative indices 47–8, 49
negative relationships 102, 105
negatively skewed distributions 119, 120
newtons 29–30, 33
nil correlation 103, 105
nomenclature, units 29
nomograms 80–1, 83
non-parametric tests 146–9
normal distribution 119, 123–30, 143–6
notation 1–9, 48–9
null hypothesis (H_0) 111, 112, 113, 136–7, 139
number notation 1–9

O

observed values 134
origin, graphs 68
overlapping ranges 120, 121
pascals 32–3
'per' 31
percentages 5–9
picometres 17
pie charts 79–80, 84
Pisum sativum 52
plotting line graphs 75–8
points, graphs 68, 77
populations, describing 118–21
positive relationships 101, 102, 103, 105
positively skewed distributions 119, 120
powers 11, 44–8
prefixes, meaning 15–16
probability (p) 135, 137, 145, 146, 153
proper fractions 4
proportion 51

P

quadrats 71, 75, 84, 111

Q

rank order 111, 147

R

ratios 51–61, 134
reciprocals, calculators 11
regression, definition 112
regression line *see* line of best fit
relationships, correlation 101–10
relative size *see* scale
resolving power 19–20
rounding 9–10, 53
r_s *see* Spearman's rank correlation test
rules, equations 36–40

S

samples 118–21, 143–9
scales
    ACFOR scale 73
    Domin scale 65
    factors 59
    logarithmic 12, 18–19, 89–91
    using 17–19
scattergrams/plots 65, 66, 103, 109
scientific calculators 11–12, 49, 109–10, 130
scientific notation 48–9
shorthand symbols 28
SI *see* International System of Units
'sigma x' 115
significance levels 135, 145, 146, 151, 152, 153
significant difference 143–6
significant figures 9–10
simplification, fractions 3–4
Simpson's diversity index (D) 149–50
size
    see also scale
    magnification 20, 21, 22, 59
    symbols 16–17
Spearman's rank correlation test (r_s) 110–12, 113
species diversity index 149–50
species linkage 138, 139
spreadsheets 63
square roots 12, 39–40
squares 39–40
stage micrometers 23–5
standard deviation 96–7, 123–30
standard error 131
standard form 48–9
statistics
    averages 114–17
    chi-squared test 134–7, 139, 153
    distribution 123–31
    Mann–Whitney *U*-test 146–9
    normal distribution 119, 123–30, 143–6
    null hypothesis 111, 112, 113, 136–7, 139
    populations 118–22
    samples 118–22, 143–9
    Simpson's diversity *D*-index 149–50
    Student's *t*-test 143–6
straight lines 78, 95
Student's *t*-test 143–6

substitution, equations 37
surface area:volume ratios 58–9
symbols 15–35
    mathematical 28, 29
    size 16–17
    units 29, 30
    using 19–28

T

t-test 143–6
tables 63, 65, 69, 82, 90
    contingency tables 138–9
    use 12–13
tally charts 123, 128, 130
ten, powers of 45–8
tendencies 103

test of association 139, 140
transects 64

U

U-test 146–9
units 15–35
    arbitrary units 82
    capital letters 29–30
    changing 28–9
    derived 31–2
    length 17
    symbols 29, 30
    using 19–28
unknowns 36
unmatched samples 145

V

variability 118
variables 38, 65
    dependent 76, 103
    discontinuous 70
    independent 67, 76, 103
volume:surface area ratios 58–9

W

water potential 32
word roots 15–16
workings, showing 8

Acknowledgements

The authors and publisher would like to thank the following for supplying photographs:

Action Plus: p. 116 (Glyn Kirk)
Ardea: p. 72 (Ian Beames), 128 (Jack A Bailey);
Getty Images: p. 18, Fig. 3, Kaz Mori (Image Bank)
Griffin: p. 22, Fig. 7
Science Photolibrary: pp. 20 (Marilyn Schaller), 22, Fig. 6 (Eye of Science), 54 (Dr Jeremy Burgess)

Thanks are also due to the following awarding bodies for kind permission to reproduce examination questions:

Assessment and Qualifications Alliance (AQA)
Edexcel
Northern Ireland Council for the Curriculum Examinations and Assessment (CCEA)